RNA-seq Data Analysis

A Practical Approach

CHAPMAN & HALL/CRC
Mathematical and Computational Biology Series

Aims and scope:
This series aims to capture new developments and summarize what is known over the entire spectrum of mathematical and computational biology and medicine. It seeks to encourage the integration of mathematical, statistical, and computational methods into biology by publishing a broad range of textbooks, reference works, and handbooks. The titles included in the series are meant to appeal to students, researchers, and professionals in the mathematical, statistical and computational sciences, fundamental biology and bioengineering, as well as interdisciplinary researchers involved in the field. The inclusion of concrete examples and applications, and programming techniques and examples, is highly encouraged.

Series Editors

N. F. Britton
Department of Mathematical Sciences
University of Bath

Xihong Lin
Department of Biostatistics
Harvard University

Hershel M. Safer
School of Computer Science
Tel Aviv University

Maria Victoria Schneider
European Bioinformatics Institute

Mona Singh
Department of Computer Science
Princeton University

Anna Tramontano
Department of Physics
University of Rome La Sapienza

Proposals for the series should be submitted to one of the series editors above or directly to:
CRC Press, Taylor & Francis Group
3 Park Square, Milton Park
Abingdon, Oxfordshire OX14 4RN
UK

Published Titles

An Introduction to Systems Biology: Design Principles of Biological Circuits
Uri Alon

Glycome Informatics: Methods and Applications
Kiyoko F. Aoki-Kinoshita

Computational Systems Biology of Cancer
Emmanuel Barillot, Laurence Calzone, Philippe Hupé, Jean-Philippe Vert, and Andrei Zinovyev

Python for Bioinformatics
Sebastian Bassi

Quantitative Biology: From Molecular to Cellular Systems
Sebastian Bassi

Methods in Medical Informatics: Fundamentals of Healthcare Programming in Perl, Python, and Ruby
Jules J. Berman

Computational Biology: A Statistical Mechanics Perspective
Ralf Blossey

Game-Theoretical Models in Biology
Mark Broom and Jan Rychtář

Computational and Visualization Techniques for Structural Bioinformatics Using Chimera
Forbes J. Burkowski

Structural Bioinformatics: An Algorithmic Approach
Forbes J. Burkowski

Spatial Ecology
Stephen Cantrell, Chris Cosner, and Shigui Ruan

Cell Mechanics: From Single Scale-Based Models to Multiscale Modeling
Arnaud Chauvière, Luigi Preziosi, and Claude Verdier

Bayesian Phylogenetics: Methods, Algorithms, and Applications
Ming-Hui Chen, Lynn Kuo, and Paul O. Lewis

Statistical Methods for QTL Mapping
Zehua Chen

Normal Mode Analysis: Theory and Applications to Biological and Chemical Systems
Qiang Cui and Ivet Bahar

Kinetic Modelling in Systems Biology
Oleg Demin and Igor Goryanin

Data Analysis Tools for DNA Microarrays
Sorin Draghici

Statistics and Data Analysis for Microarrays Using R and Bioconductor, Second Edition
Sorin Drăghici

Computational Neuroscience: A Comprehensive Approach
Jianfeng Feng

Biological Sequence Analysis Using the SeqAn C++ Library
Andreas Gogol-Döring and Knut Reinert

Gene Expression Studies Using Affymetrix Microarrays
Hinrich Göhlmann and Willem Talloen

Handbook of Hidden Markov Models in Bioinformatics
Martin Gollery

Meta-analysis and Combining Information in Genetics and Genomics
Rudy Guerra and Darlene R. Goldstein

Differential Equations and Mathematical Biology, Second Edition
D.S. Jones, M.J. Plank, and B.D. Sleeman

Knowledge Discovery in Proteomics
Igor Jurisica and Dennis Wigle

Introduction to Proteins: Structure, Function, and Motion
Amit Kessel and Nir Ben-Tal

RNA-seq Data Analysis: A Practical Approach
Eija Korpelainen, Jarno Tuimala, Panu Somervuo, Mikael Huss, and Garry Wong

Biological Computation
Ehud Lamm and Ron Unger

Optimal Control Applied to Biological Models
Suzanne Lenhart and John T. Workman

Published Titles (continued)

Clustering in Bioinformatics and Drug Discovery
John D. MacCuish and Norah E. MacCuish

Spatiotemporal Patterns in Ecology and Epidemiology: Theory, Models, and Simulation
Horst Malchow, Sergei V. Petrovskii, and Ezio Venturino

Stochastic Dynamics for Systems Biology
Christian Mazza and Michel Benaïm

Engineering Genetic Circuits
Chris J. Myers

Pattern Discovery in Bioinformatics: Theory & Algorithms
Laxmi Parida

Exactly Solvable Models of Biological Invasion
Sergei V. Petrovskii and Bai-Lian Li

Computational Hydrodynamics of Capsules and Biological Cells
C. Pozrikidis

Modeling and Simulation of Capsules and Biological Cells
C. Pozrikidis

Cancer Modelling and Simulation
Luigi Preziosi

Introduction to Bio-Ontologies
Peter N. Robinson and Sebastian Bauer

Dynamics of Biological Systems
Michael Small

Genome Annotation
Jung Soh, Paul M.K. Gordon, and Christoph W. Sensen

Niche Modeling: Predictions from Statistical Distributions
David Stockwell

Algorithms in Bioinformatics: A Practical Introduction
Wing-Kin Sung

Introduction to Bioinformatics
Anna Tramontano

The Ten Most Wanted Solutions in Protein Bioinformatics
Anna Tramontano

Combinatorial Pattern Matching Algorithms in Computational Biology Using Perl and R
Gabriel Valiente

Managing Your Biological Data with Python
Allegra Via, Kristian Rother, and Anna Tramontano

Cancer Systems Biology
Edwin Wang

Stochastic Modelling for Systems Biology, Second Edition
Darren J. Wilkinson

Bioinformatics: A Practical Approach
Shui Qing Ye

Introduction to Computational Proteomics
Golan Yona

Chapman & Hall/CRC Mathematical and Computational Biology Series

RNA-seq Data Analysis

A Practical Approach

Eija Korpelainen
CSC - IT Center for Science
Espoo, Finland

Jarno Tuimala
RS-koulutus
Helsinki, Finland

Panu Somervuo
University of Helsinki
Finland

Mikael Huss
SciLifeLab, Stockholm University
Sweden

Garry Wong
University of Eastern Finland
Kuopio, Finland

CRC Press is an imprint of the
Taylor & Francis Group, an **informa** business
A CHAPMAN & HALL BOOK

Chapman & Hall/CRC Press
Taylor & Francis Group
6000 Broken Sound Parkway NW, Suite 300
Boca Raton, FL 33487-2742

© 2015 by Taylor & Francis Group, LLC
Chapman & Hall/CRC Press is an imprint of Taylor & Francis Group, an Informa business

No claim to original U.S. Government works

Version Date: 20140630

ISBN 13: 978-1-4665-9500-2 (hbk)

This book contains information obtained from authentic and highly regarded sources. Reasonable efforts have been made to publish reliable data and information, but the author and publisher cannot assume responsibility for the validity of all materials or the consequences of their use. The authors and publishers have attempted to trace the copyright holders of all material reproduced in this publication and apologize to copyright holders if permission to publish in this form has not been obtained. If any copyright material has not been acknowledged please write and let us know so we may rectify in any future reprint.

Except as permitted under U.S. Copyright Law, no part of this book may be reprinted, reproduced, transmitted, or utilized in any form by any electronic, mechanical, or other means, now known or hereafter invented, including photocopying, microfilming, and recording, or in any information storage or retrieval system, without written permission from the publishers.

For permission to photocopy or use material electronically from this work, please access www.copyright.com (http://www.copyright.com/) or contact the Copyright Clearance Center, Inc. (CCC), 222 Rosewood Drive, Danvers, MA 01923, 978-750-8400. CCC is a not-for-profit organization that provides licenses and registration for a variety of users. For organizations that have been granted a photocopy license by the CCC, a separate system of payment has been arranged.

Trademark Notice: Product or corporate names may be trademarks or registered trademarks, and are used only for identification and explanation without intent to infringe.

Library of Congress Cataloging-in-Publication Data

Korpelainen, Eija, author.
RNA-seq data analysis : a practical approach / Eija Korpelainen, Jarno Tuimala, Panu Somervuo, Mikael Huss, Garry Wong.
p. ; cm.
Includes bibliographical references and index.
ISBN 978-1-4665-9500-2 (hardcover : alk. paper)
I. Tuimala, Jarno, author. II. Somervuo, Panu, author. III. Huss, Mikael, author. IV. Wong, Garry, author. V. Title.
[DNLM: 1. Sequence Analysis, RNA--methods. 2. Transcriptome. 3. Statistics as Topic. QU 58.7]

QP623
572.8'8--dc23 2014024218

Visit the Taylor & Francis Web site at
http://www.taylorandfrancis.com

and the CRC Press Web site at
http://www.crcpress.com

Printed and bound by CPI Group (UK) Ltd, Croydon, CR0 4YY

Contents

Preface

A PRACTICAL BOOK FOR VARIOUS AUDIENCES

RNA-seq offers unprecedented information about the transcriptome, but harnessing this information with bioinformatic tools is typically a bottleneck. The goal of this book is to enable readers to analyze RNA sequencing (RNA-seq) data. Several topics are discussed in detail, covering the whole data analysis workflow from quality control, mapping, and assembly to statistical testing and pathway analysis. Instead of minimizing overlap with existing textbooks, the aim is for a more comprehensive and practical presentation.

This book enables researchers to examine differential expression at gene, exon, and transcript levels and to discover novel genes, transcripts, and whole transcriptomes. In keeping with the important regulatory role of small noncoding RNAs, a whole section is devoted to their discovery and functional analysis.

This book is intended for students and advanced researchers alike. Practical examples are chosen in such a way that not only bioinformaticians, but also nonprogramming wet lab scientists can follow them, making the book suitable for researchers from a wide variety of backgrounds including biology, medicine, genetics, and computer science. This book can be used as a textbook for a graduate course, for an advanced undergraduate course, or for a summer school. It can serve as a self-contained handbook and detailed overview of the central RNA-seq data analysis methods and how to use these in practice.

The book balances theory with practice, so that each chapter starts with theoretical background, followed by descriptions of relevant analysis tools, and examples of their usage. In line with our desire for accessibility, the book is a self-contained guide to RNA-seq data analysis. Importantly, it caters also to noncomputer-savvy wet lab biologists, as examples are given using the graphical Chipster software in addition to command line tools. All software used in examples are open source and freely available.

OUTLINE OF THE CONTENTS

The "Introduction" section in Chapters 1 and 2 discusses the different applications of RNA-seq, ranging from discovery of genes and transcripts to differential expression analysis and discovery of mutations and fusion genes. It gives an overview of RNA-seq data analysis and discusses important aspects of experimental planning.

The first part is devoted to mapping reads to references and *de novo* assembly. As the quality of reads strongly influences both, a chapter on quality control and preprocessing is also included. Chapter 3 discusses several quality issues characteristic to high-throughput sequencing data and tools for detecting and solving them. Chapter 4 describes the challenges in mapping RNA-seq reads to a reference and introduces some commonly used aligners with practical examples. Tools for manipulating alignment files and genome browsers for visualizing reads in genomic context are introduced. Chapter 5 describes the elements of transcriptome assembly. The relevance of data-processing steps such as filtering, trimming, and error correction in RNA-seq assembly is discussed. Fundamental concepts such as splicing graph, de Bruijn graph, and path traversal in an assembly graph are explained. Differences between genome and transcriptome assembly are discussed. Two approaches to reconstruct full-length transcripts are explained: mapping-based and *de novo* assembly. Both approaches are also demonstrated with practical examples.

The second part is devoted to statistical analyses predominantly carried out with the R software that is supplemented with the tools produced by the Bioconductor project. Chapter 6 discusses the different quantitation approaches and tools, as well as annotation-based quality metrics. Chapter 7 describes the R- and Bioconductor-based framework for the analysis of RNA-seq data and how to import the data in R. The main differences between the statistical and bioinformatic tools in R are also discussed. Chapters 8 and 9 discuss different options for analyzing differential expression of genes, transcripts, and exons and show how to perform the analysis using R/Bioconductor tools and some standalone tools. Chapter 10 offers solutions for annotating results, and Chapter 11 describes different ways of producing informative visualizations to display the central results.

The last part of the book is focused on the analysis of small noncoding RNAs using web-based or free downloadable tools. Chapter 12 describes the different classes of small noncoding RNAs and characterizes their function, abundance, and sequence attributes. Chapter 13 describes different algorithms that are used to discover small noncoding RNAs from next-generation sequencing data sets and provides a practical approach with workflows and examples of how small noncoding RNAs are discovered and annotated. In addition, the chapter describes downstream tools that can be used to elucidate functions of small noncoding RNAs.

Acknowledgments

We thank the staff at CRC Press for giving us this opportunity to write a textbook for the RNA-seq community. In particular, Sunil Nair, Sarah Gelson, and Stephanie Morkert guided us during the writing process, demonstrated limitless patience, and were quick and prompt with every query we had.

We also thank colleagues and members of our laboratories for reading and commenting on chapters: Vuokko Aarnio, Liisa Heikkinen, and Juhani Peltonen. Their time and efforts are deeply appreciated.

Tommy Kissa provided unwavering and unconditional enthusiasm as an assistant during the final writing stages of this work. His attitude that this was the most enjoyable task of the day provided inspiration and was not lost on the final results.

Finally, we thank our spouses and family members Lily, Philippe, Stefan, Sanna, and Merja who acted as research assistants, reviewers, graphic artists, computer support, and sounding boards, in addition to serving as hoteliers, caterers, maids, and psychological therapists from the first to last word. We lovingly dedicate this work to you.

Authors

Eija Korpelainen works as a bioinformatician at the CSC-IT Center for Science. She has over a decade of experience in providing bioinformatics support at a national level. Her team developed the Chipster software which offers a user-friendly interface to a comprehensive collection of analysis and visualization tools for microarray and next-generation sequencing data. She also runs several training courses both in Finland and abroad.

Jarno Tuimala has worked as a biostatistician and bioinformatician at the Finnish Red Cross Blood Service and CSC-IT Center for Science. He is also an adjunct professor of bioinformatics at the University of Helsinki. He has over a decade of experience working with the R software and analyzing high-throughput data.

Panu Somervuo earned his D.Sc. (Tech.) degree from Helsinki University of Technology, Finland, in 2000. His background is in signal processing, pattern recognition, and machine learning. He worked with automatic speech recognition and neural networks before moving into bioinformatics. During the past 6 years, he has been involved in several projects utilizing microarrays and next-generation sequencing at the University of Helsinki.

Mikael Huss has worked with bioinformatics for high-throughput sequencing since 2007 as a postdoc at the Genome Institute of Singapore, in the sequencing facility at SciLifeLab in Stockholm (where he designed the RNA-seq analysis workflows) and in his current position as an "embedded bioinformatician" in SciLifeLab's WABI, a national resource for providing project-oriented bioinformatics analysis support. He has over a decade's experience in computational biology.

Garry Wong held the professorship in molecular bioinformatics at the University of Eastern Finland and is currently Professor of Biomedical Sciences at the University of Macau. He has over a decade of experience in using and developing transcriptomic tools for analysis of biologically active RNA molecules. His laboratory is currently focused on elucidating the functions of noncoding small RNAs in model organisms using bioinformatic and functional genomic tools.

CHAPTER 1

Introduction to RNA-seq

1.1 INTRODUCTION

RNA-seq describes a collection of experimental and computational methods to determine the identity and abundance of RNA sequences in biological samples. Thus, the order of each adenosine, cytosine, guanine, and uracil ribonucleic acid residue present in a single-stranded RNA molecule is identified. The experimental methods involve isolation of RNA from cell, tissue, or whole-animal samples, preparation of libraries that represent RNA species in the samples, actual chemical sequencing of the library, and subsequent bioinformatic data analysis. A critical distinction of RNA-seq from earlier methods, such as microarrays, is the incredibly high throughput of current RNA-seq platforms, the sensitivity afforded by newer technologies, and the ability to discover novel transcripts, gene models, and small noncoding RNA species.

RNA-seq methods are derived from generational changes in sequencing technology. First-generation high-throughput sequencing typically refers to Sanger dideoxy sequencing. With capillary electrophoresis being utilized to resolve nucleic acid fragment lengths, a standard run might employ 96 capillaries and generate a sequence length of 600–1000 bases yielding approximately 100,000 bases of sequence. Second-generation sequencing, also known as next-generation sequencing (NGS), refers to methods using similar sequencing by synthesis chemistry of individual nucleotides, but performed in a massively parallel format, so that the number of sequencing reactions in a single run can be in millions. A typical NGS run could consist of 6000 M sequencing reactions of 100 nucleotides yielding 600 billion bases of sequence information.

Third-generation sequencing refers to methods that are also massively parallel and use sequence by synthesis chemistry but have as templates individual molecules of DNA or RNA. Third-generation sequencing platforms have fewer sequencing reactions per run, in the order of a few millions, but the length of sequence per reaction can be larger and can easily run into the 1500 nucleotide range.

Data obtained from an RNA-seq experiment can produce new knowledge ranging from the identification of novel protein coding transcripts in embryonic stem cells to characterization of over-expressed transcripts in skin cancer cell lines. Questions that can be asked include: What are the differences in the levels of gene expression in normal and cancer cells? What happens to the gene expression levels in cell lines missing a tumor suppressor gene? What are the gene expression differences in my cell line before and after mutagen treatment? Which genes are up-regulated during the development of brain? What transcripts are present in skin but not in muscle? How is gene splicing changed during oxidative stress? What novel miRNAs can we discover in a human embryonic stem-cell sample? As one can see, the range of questions that can be asked is broad.

Excitement and heightened expectations for transcriptomics arrived when RNA-seq technologies revealed that the current knowledge of gene structure and the general annotation of genes, from single-cell model organisms to human cells, was quite poor. New data derived from RNA-seq platforms showed a vast diversity for gene structure, identified novel unknown genes, and shed light on noncoding transcripts of both small and long lengths [1–4]. Later studies provided massive amounts of data for many new species that had very limited transcript sequence information available. The pace of research has been such that a well-known analogy in the sequencing community is that the cost of sequencing is declining at a rate faster than Moore's law. With such advantageous economics come fantastic gains in productivity and even greater expectations.

This book focuses on practical approaches in data analysis methods in RNA-seq; however, it would be impossible to describe these approaches without presenting the experimental methods. In this introductory chapter, we will provide some necessary background, show some typical protocols, provide a workflow, and finally provide some examples of successful applications. The reader will then hopefully have a better understanding of the entire process, step by step, from conceptualization of the project to visualization and interpretation of the results.

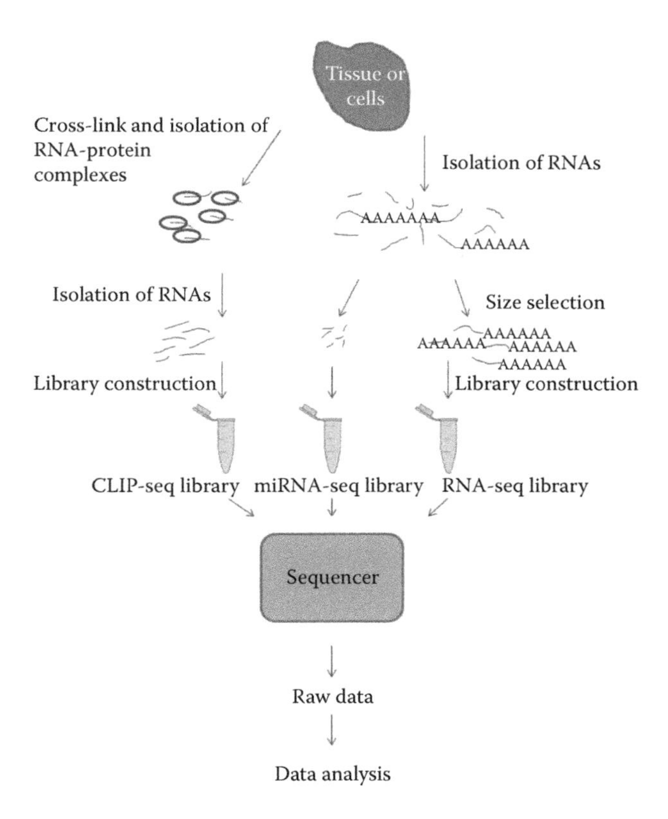

FIGURE 1.1 General scheme for RNA-seq experiments. The workflow from tissue to data in the RNA-seq method is shown with alternatives for CLIP-seq, miRNA-seq, and general RNA-seq.

A typical workflow for RNA-seq is shown in Figure 1.1. The beginning of the workflow shows wet-laboratory steps, whereas the bottom shows the data handling and analysis steps.

1.2 ISOLATION OF RNAs

RNAs are typically isolated from freshly dissected or frozen cells or tissue samples using commercially available kits such as RNAEasy (Qiagen Hilden, Germany), TRIZOL (Life Technologies, Carlsbad, CA), or RiboPure (Ambion, Austin, TX), among many others. These kits have the advantage of being easy to use and yielding large amounts of total RNA when used properly. High-throughput RNA isolation systems also exist that relies mainly on RNA attached to magnetic particles which facilitate their washing and isolation. It is also possible, although not ideal, to isolate RNA from formalin-fixed, paraffin-embedded tissues. To prevent degradation of RNA, samples can be immersed in RNA storage reagents such as RNAlater (Ambion), or processed partially and stored as a phenolic emulsion (Trizol). At this stage, RNA samples can also be enriched for size-specific classes such as small RNAs using column systems (miRVana; Ambion). Alternatively, samples can be isolated initially as total RNA and then size selected by polyacrylamide gel electrophoresis.

In almost all cases of total RNA isolation, the sample will be contaminated by genomic DNA. This is unavoidable, and even if the contamination is minor, the sensitivity and throughput of RNA-seq will eventually capture these contaminants. Therefore, it is common practice that total RNA-isolated samples are treated with DNase, to digest contaminating DNA prior to library preparation. Most DNase kits provide reagents for inactivating DNase once the offending DNA has been removed. The amount of total RNA required for RNA-seq library preparation varies. Standard library protocols require 0.1–10 μg of total RNA, and high-sensitivity protocols can produce libraries from as little as 10 pg of RNA. It is becoming common that RNA from single cells is isolated and specific kits for these applications are becoming available.

1.3 QUALITY CONTROL OF RNA

Best practices require that RNAs are quality checked for degradation, purity, and quantity prior to library preparation. Several platforms are available for this step. Nanodrop and similar devices measure the fluorescent absorbance of nucleic acid samples typically at 260 and 280 nm. It requires only a fraction of a microliter of liquid for measurement, in which a sample can be diluted, thus using nanoliter amounts of starting material. The device is very easy to use, takes seconds to obtain a reading, and can handle many samples simultaneously. As the device measures absorbance of the sample, it is not able to distinguish between RNA and DNA, and therefore cannot indicate whether the RNA sample is contaminated with DNA. Moreover, degraded RNA will give similar readings as intact RNA, and therefore we cannot know about the quality of the sample. The 260/280 absorbance ratio will, however, provide some information about contamination by proteins.

As pipetting samples in the nanoliter range is at the limit of common laboratory pipettors, accuracy of measurement at the lowest concentration ranges (ng/μL) may be challenging. QubitFluorometer (Life Technologies) and similar systems that measure fluorescence of nucleic acid-derivatized products measure more directly either RNA or DNA in samples. Using measurements of low-concentration standards coupled with placement of fluorescent values on a calibration regression line of standards, and subsequent plotting of the sample fluorescent measurements on the regression line, more specific, accurate, and wider dynamic range measurement of RNAs is achievable. In addition, small volumes of less than a microliter are sufficient for measurement and even these can be diluted. While being simple to use, these systems still do not provide any measure of degradation. To deal with this problem, another instrument needs to be used.

Agilent Bioanalyzer is a microfluidics capillary electrophoresis-based system to measure nucleic acids. It combines the advantage of small volumes and sensitivity, with electrophoresis being used for sizing nucleic acid samples. When size standards are run, the sizing and quantitation of RNAs in the sample provides critical information not only on the concentration, but also on the quality of nucleic acid. Degraded RNAs will appear as a smear at low-molecular weights, whereas intact total RNA will show sharp 28S and 18S peaks. The Bioanalyzer system contains a microchip that is loaded with size controls and space for up to 12 samples at a time. Samples are mixed with a polymer and a fluorescent dye, which are then loaded and measured through capillary electrophoretic movement. The integrated data analysis pipeline on the instrument will also render the electrophoretic data into a gel-like picture for users more accustomed to traditional gel electrophoresis. A sample of a Bioanalyzer run is shown in Figure 1.2.

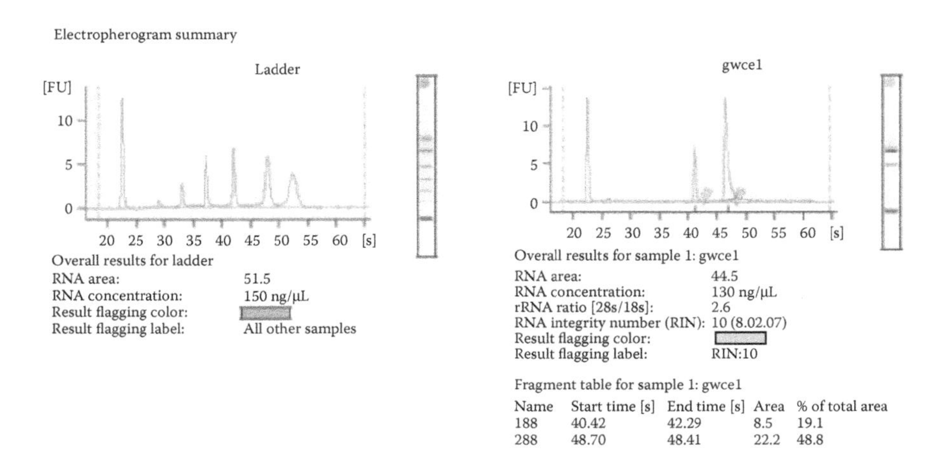

FIGURE 1.2 Agilent Bioanalyzer run showing RNA quality. Both the ladder and sample run are shown.

1.4 LIBRARY PREPARATION

Prior to sequencing, the RNAs in a sample are converted into a cDNA library representing all of the RNA molecules in the sample. This step is performed because in practice, RNA molecules are not directly sequenced, instead DNAs are sequenced due to their better chemical stability, and are also more amenable to the sequencing chemistry and protocols of each sequencing platform. Therefore, the library preparation has two purposes, the first is to faithfully represent the RNAs in the sample and secondly to convert RNA into DNA. Each RNA-sequencing platform (e.g., Illumina, Solid, Ion Torrent) has its own specific protocol, so there is no need for providing separate protocols for each. The library protocols for each commercial platform are available with their kits at the company's website (Table 1.1).

Third-party library preparation kits are also available and are being used successfully. It is also possible to create one's own kit using commonly available molecular biology components although this lacks the convenience, optimization, and support of commercial products. Here, we show the typical library protocol steps for the Illumina platform for RNA-seq. A schematic of the steps is shown in Figure 1.3.

The major steps in library preparation involves the following:

1. Obtain pure, intact, and quality-checked total RNA of approximately 1–10 μg. The exact amount needed depends on the application and platform.

TABLE 1.1 Major RNA-seq Platforms and Their General Properties

Platform	Sequencing Chemistry	Detection Chemistry	Weblink
Illumina	Sequencing by synthesis	Fluorescence	www.illumina.com
SOLID	Sequencing by ligation	Fluorescence	www.invitrogen.com
Roche 454	Sequencing by synthesis	Luminescence	www.454.com
Ion Torrent	Sequencing by synthesis	Proton release	www.iontorrent.com
Pacific biosciences	Single-molecule seq by synthesis	Real-time fluorescence	www.pacificbiosciences.com
Oxford nanopore	Single-molecule seq by synthesis	Electrical current difference per nucleotide through a pore	www.nanoporetech.com

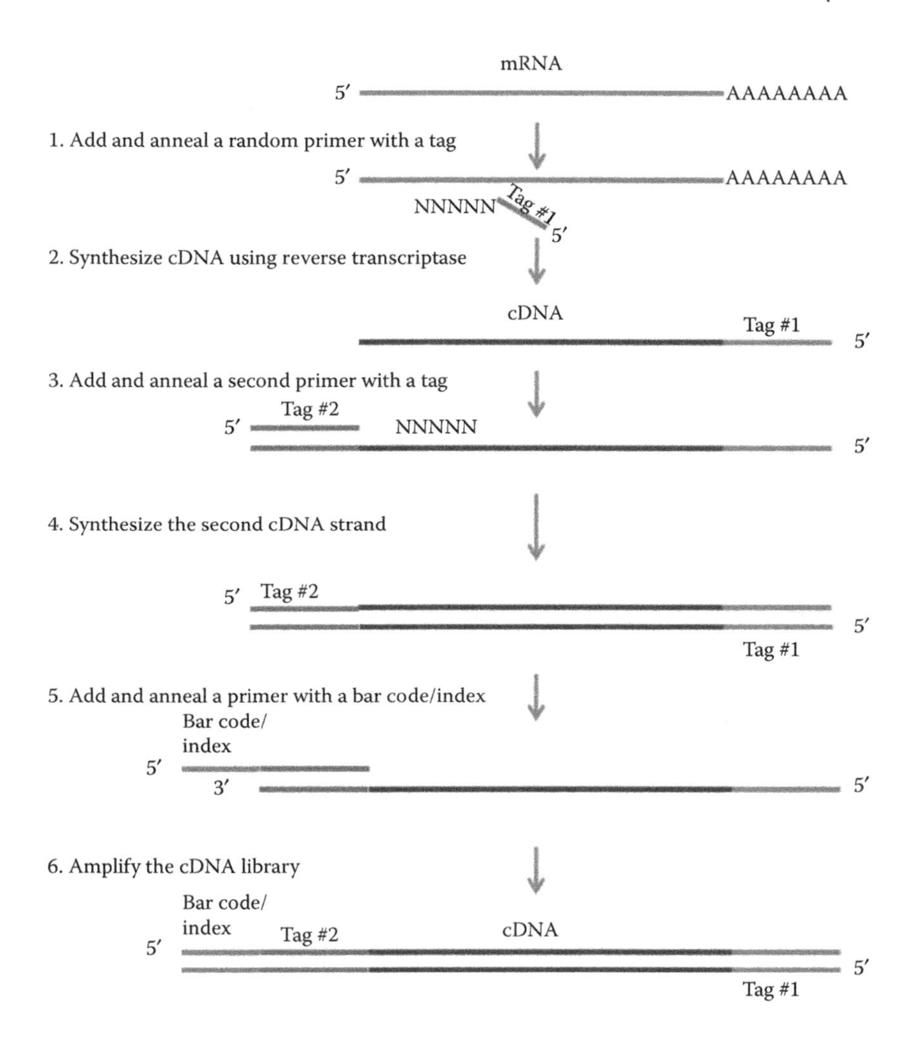

FIGURE 1.3 Schematic of RNA-seq library preparation.

2. Purify mRNA from the total RNA. Typically, this is done by annealing total RNA to oligo-dT magnetic beads. Two rounds of purification may be performed to remove nonspecifically bound ribosomal and other RNAs from the oligo-dT. Release or dissociate mRNAs from oligo-dT beads.

3. Fragment purify mRNA by incubation with fragmentation reagent. This breaks the mRNA strands into multiple small fragments.

4. Prime the fragmented mRNAs with random hexamer primers.

5. Reverse-transcribe the fragmented mRNAs with Reverse Transcriptase, thus producing cDNAs.

6. Synthesize the second/opposite strand of the cDNA and remove the RNA. The product will be double-stranded cDNA (ds cDNA).

7. Purify the ds cDNA from free nucleotides, enzymes, buffers, and RNA. This can be done by binding the DNA with Solid-Phase Reversible Immobilization (SPRI) beads, for example. The advantage of using these paramagnetic beads is that once bound, the beads can be washed to purify the ds cDNA which remains on the beads. Once washed, the ds cDNA can be eluted from the beads for the next reaction.

8. Perform end-repair on purified eluted ds cDNA.

9. Purify the end-repaired ds cDNA. This can also be done on SPRI beads.

10. Adenylate 3′ ends of eluted end-repaired ds cDNA.

11. Ligate adaptors to the end-repaired ds cDNA. Adaptors will ligate to both ends of the ds cDNA. These adaptors can be indexed for each library reaction. In other words, each adaptor can have a six-nucleotide difference in the adaptor sequence. Using a different index for each library reaction allows for pooling libraries later for sequencing, yet still allowing for tracing the sequence back to the original library based on the adaptor sequence.

12. Purify the adaptor ligated, end-repaired ds cDNA. Again, this can be done with SPRI beads.

13. Enrich the library by polymerase chain reaction (PCR) amplification. Using sequences from the adaptor as primers, small numbers of cycles (12–16) are used to amplify the sequences already present.

14. Purify the PCR-enriched, adaptor-ligated, end-repaired ds cDNA. Again, this can be done with SPRI beads. This is now the library representing the original mRNAs in the sample.

15. Validate and quality-control the library. This can be done in several ways by (1) selectively amplifying via PCR-specific genes that should be present in the library; (2) quantifying the yield of ds cDNA in the library; (3) visualizing the abundance and size distribution of the

library by polyacrylamide gel electrophoresis, or capillary electrophoresis on an Agilent Bioanalyzer.

16. Normalize and pool libraries. As the capacity to sequence in a single flow cell is enormous, it is possible to sequence many libraries (up to 24 libraries/flow cell lane is possible, 6–12 is of more normal practice). Normalization acts to even out the amounts of ds cDNA in each library. For example, all libraries can be diluted to 10 nM ds cDNA and then pooled at even volumes, so that each library is equally represented.

17. Send normalized and pooled libraries to sequencing facility for cluster generation and sequencing protocol which is dependent on the specific platform (Illumina, Solid, 454, etc.).

1.5 MAJOR RNA-SEQ PLATFORMS

1.5.1 Illumina

This platform represents one of the most popularly used sequencing by synthesis chemistry in a massively parallel arrangement. After libraries are made, ds cDNA is passed through a flow cell which will hybridize the individual molecules based on complementarity with adaptor sequences. Hybridized sequences held at both ends of the adaptor by the flow cell will be amplified as a bridge. These newly generated sequences will hybridize to the flow cell close by and after many cycles a region of the flow cell will contain many copies of the original ds cDNA. This entire process is known as cluster generation. After the clusters are generated, and one strand removed from the ds cDNA, reagents are passed through the flow cell to execute sequencing by synthesis. Sequencing by synthesis describes a reaction where in each synthesis round, the addition of a single nucleotide, which can be A, C, G, or T, as determined by a fluorescent signal, is imaged, so that the location and added nucleotide can be determined, stored, and analyzed. Reconstruction of the sequence of additions in a specific location on the flow cell, which corresponds to a generated ds cDNA cluster, gives the precise nucleotide sequence for an original piece of ds cDNA.

The number of synthesis rounds can be less, for example, from 50 nucleotides (nt) to 150 nt. There are also two modes in which sequencing can be performed. If sequencing is performed at one end of the ds cDNA only, it is single read mode. If sequencing is performed from both ends, it is termed paired-end read mode. Abbreviations for the type of read and length are typically SR50 or PE100, indicating single read 50 nt or paired-end read 100 nt, respectively. Since each cycle requires washing of used reagents and introduction of new reagents, a single sequencing run on the instrument may take anywhere from 3 to 12 days depending upon the instrument model and sequencing length. Illumina provides a wide range of instruments with different throughputs. The Hi-Seq 2500 instrument produces up to 6 billion paired-end reads in a single run. At PE100, this represents 600 Gb of data. This is massively more sequence data than is typically needed for a single study, so that, in practice, the libraries are indexed and several libraries are normalized and combined to be run on a single flow cell. It is normal practice to have as many as a hundred libraries run in total on a 16-lane flow cell. If this is too much sequencing capacity for a laboratory, Illumina also provides a smaller, yet more personal sequencer with lower throughput. The MiSeq system can produce 30 M reads in PE250 mode representing 8.5 Gb of data within a 2-day runtime.

1.5.2 SOLID

SOLID stands for sequencing by oligonucleotide ligation and detection and is a platform commercialized by Applied Biosystems (Carlsbad, CA). As the name implies, the sequencing chemistry is via ligation rather than synthesis. In the SOLID platform, a library of DNA fragments (originally derived from RNA molecules) is attached to magnetic beads at one molecule per bead. The DNA on each bead is then amplified in an emulsion so that amplified products remain with the bead. The resulting amplified products are then covalently bound to a glass slide. Using several primers that hybridize to a universal primer, di-base probes with fluorescent labels are competitively ligated to the primer. If the bases in the first and second positions of the di-base probe are complementary to the sequence, then the ligation reaction will occur and the label will provide a signal. Primers are reset by a single nucleotide five times, so at the end of the cycle, at least four nucleotides would have been interrogated twice due to the dinucleotide probes and the fifth nucleotide at least once. Ligation of subsequent dinucleotide probes provides a second interrogation of the fifth nucleotide and after five primer resets, five more nucleotides will have been interrogated at least twice. The ligation steps continue until the sequence is read.

The unique ligation chemistry allows for two checks of a nucleotide position and thus provides greater sequencing accuracy of up to 99.99%. While this may not be necessary for applications such as differential expression, it is critical for detecting single-nucleotide polymorphisms

(SNPs). The newest instruments such as the 5500 W do away with bead amplification and use flow chips in place of amplifying templates. The throughput can be up to 320 Gb of data from two flow chips. As with other platforms, indexing/barcoding can be used to multiplex libraries so that hundreds of library samples can be run simultaneously on the instrument.

1.5.3 Roche 454

This platform is also based on adaptor-ligated ds DNA library sequencing by synthesis chemistry. ds DNA is fixed onto beads and amplified in a water–oil emulsion. The beads are then placed into picotiter plates where sequencing reactions take place. The massive numbers of wells in picotiter plates provide the massively parallel layout needed for NGS. The detection method differs from other platforms in that the synthesis chemistry involves detection of an added nucleotide via a two-step reaction.

The first step cleaves the triphosphate nucleotide after an addition, releasing pyrophosphate. The second step converts pyrophosphate into adenosine triphosphate (ATP) via the enzyme ATP sulfurylase. The third step uses the newly synthesized ATP to catalyze the conversion of luciferin into oxyluciferin via the enzyme luciferase and this reaction generates a quanta of light that is captured from the picotiter plate by a charge-coupled camera. Free nucleotides and unreacted ATP are degraded by a pyrase after each addition. These steps are repeated until a predetermined number of reactions have been reached. Recording the light generation and well location after each nucleotide addition allows for reconstruction of the identity of the nucleotide and the sequence for each well.

This method is termed pyrosequencing and the advantage of this sequencing chemistry is that it permits for longer reads when compared to other platforms. Read lengths of up to 1000 bases can be achieved on this platform. Roche owns this platform and provides the current GS FLX+ system as well as a smaller GS junior system. With up to 1 million reads per run, and an average of 700 nt per read, 700 Mb of sequence data can be achieved in less than 1 day of run time.

1.5.4 Ion Torrent

This newer platform utilizes the adaptor-ligated library followed by sequencing-by-synthesis chemistry of other platforms, but has a unique feature. Instead of detecting fluorescent signals or photons, it detects changes in the pH of the solution in a well when a nucleotide is added and protons are produced. These changes are miniscule, however the Ion Torrent device utilizes technologies developed in the semiconductor industry to achieve detectors of sufficient sensitivity and scales that are useful for nucleic acid sequencing. One limitation that has been pointed out is that homopolymers may be difficult to read as there is no way to stop the addition of only one nucleotide if the same nucleotide is next in the sequence. Ion Torrent can detect a larger change in the pH and uses this measurement to read through polymer regions.

This platform produces overall fewer reads than the others in a single run. For example, 60–80 M reads at 200 bases per read are possible on the Proton instrument in a run producing 10 Gb of data. However, the run time is only 2–4 h instead of 1–2 weeks on other platforms. Since neither modified nucleotides nor optical measurement instrumentation is needed, an advantage of this platform is affordability, both of the instrument and reagents. The machine has a small footprint, can be powered down when not in use and easily brought back to use, and requires minimal maintenance. With the convenience, size, and speed, it has found sizable applications in microbe sequencing, environmental genomics, and clinical applications where time is critical. This platform is also very popular for amplicon sequencing and use of primer panels for amplicon sequencing developed by specific user communities. Its low-cost and small footprint have also made it attractive to laboratories wishing to have their own personal sequencer.

1.5.5 Pacific Biosciences

This is a platform representative of the third generation. The chemistry is still similar to second generation as it is a sequencing-by-synthesis system; however, a major difference is that it requires only a single molecule and reads the added nucleotides in real time. Thus, the chemistry has been termed SMRT for single-molecule real time. Single-molecule chemistry means that no amplification needs to be performed. It has to be borne in mind that this platform sequences DNA molecules.

SMRT, as implemented by Pacific Biosciences Instruments, uses zero-mode waveguides (ZMWs) as the basis of their technology. ZMWs are space-restricted chambers that allow guidance of light energy and reagents into extremely small volumes that are in the order of zeptoliters (10^{-21} L). In the context of the Pacific Biosciences platform, this translates to a single chamber that contains a single molecule of DNA polymerase and a single

DNA molecule that is sequenced in real time. Using specific fluorescent nucleotide triphosphates, the addition of an A, C, G, or T to a nucleotide chain can be detected as it is being synthesized. The advantage of speed is enormous. As a real-time instrument that measures additions as they happen, the runtime can be very short, in the order of only one or two hours. Average read lengths can be 5000 nt. Improvements in the enzyme and synthesis chemistry can produce routine reads of up to 10,000 nt with longest reads up to 30,000 nt. The current version of the instrument called the PacBio RS II can thus produce up to 250 Mb of sequence in a single run, so even throughput is not compromised.

As a consequence of direct DNA sequencing of single molecules, it was noticed that nucleic acid modifications such as 5-methyl cytosine caused consistent and reproducible delays in the kinetics of the sequencing DNA polymerase. This has been exploited in the platform to provide sequencing of DNA modifications. Currently, detection of up to 25 base modifications is claimed to be possible on this platform.

1.5.6 Nanopore Technologies

Despite the impressive gains in throughput and low per base cost of current sequencing, efforts continue to improve sequencing technologies. While current nanopore technologies are in prototype or development, they so far have had minimal impact on RNA-seq studies. However, their impact in the future may be greater. Nanopore sequencing is a third-generation single-molecule technique where a single enzyme is used to separate a DNA strand and guides it through a protein pore embedded in a membrane. Ions simultaneously pass through the pore to generate an electric current that is measured. The current is sensitive to specific nucleotides passing through the pore, thus A, C, G, or T impede current flow differently and produce a signal that is measured in the pore. The advantage of this system is its simplicity leading to small-platform device size (e.g., early claims suggested that this would be a USB stick-sized device), but the system is technically challenging due to the need to measure very small changes in current at single-molecule scale. The efforts to commercialize this technology are led by Oxford Nanopore, however Illumina also has nanopore sequencing under development. Oxford Nanopore technologies are slated to measure directly RNA, DNA, or protein as it passes through a manufactured pore. Although this technology is not widely available at a commercial level, it shows a lot of promise.

1.6 RNA-SEQ APPLICATIONS

The main goals of RNA-seq are to identify the sequence, structure, and abundance of RNA molecules in a specific sample. By sequence, we mean the particular order of A, C, G, U residues. By structure, we mean the gene structure [i.e., location of promoter, intron–exon junctions, 5′ and 3′ untranslated regions (UTRs), and polyA site]. Secondary structure provides the locations of complementary nucleotide pairing and hairpins or bulges. Tertiary structure provides the three-dimensional shape of the molecule. By abundance, we mean the numerical amounts of each particular sequence both as absolute and normalized values. Sequence can be used to identify known protein-coding genes, novel genes, or long noncoding RNAs. Once sequence has been determined, folding into secondary structures can reveal the class of molecules such as tRNA or miRNA. Comparison of the abundance of reads for each RNA species can be made between samples derived from different developmental stages, body parts, or across closely related species. Below, we will present some common applications to provide the range of questions that can be asked and answered using RNA-seq methods. Where appropriate we also provide some examples from the scientific literature.

1.6.1 Protein Coding Gene Structure

Earlier transcriptomic methods such as cloning and Sanger sequencing of cDNA libraries, microarray expression analysis, and serial analysis of gene expression (SAGE), as well as computational prediction from genomic sequences already provide gene structures. These structures have been archived in databases and provide an easily accessible source for comparing raw RNA-seq data with known protein coding genes. As an important first step, RNA-seq reads are often initially mapped to known protein-coding genes. In addition to confirming exon–intron boundaries, the RNA-seq data can also show evidence for both shorter and longer exon boundaries, as well as the existence of completely novel exons. The collection of exons and introns that make up a gene is called a gene model. Since RNA-seq is quantitative, it can also show usage within a sample of alternative exon boundaries or alternative exons: for example, when a specific exon is used five times more often than another one. Similarly, the 5′ transcription start site (TSS) can be mapped precisely. Alternatively 5′TSS can also be identified. At the 3′ end of the molecule, the 3′UTR can be identified precisely such that the site of polyadenylation can be observed in the RNA-seq reads. Alternative polyadenylation sites can also be observed in the same way as alternative TSS as well as their respective abundances. As RNA-seq is massively parallel, sufficient reads will permit these gene structures and their alternatives to be mapped for presumably every protein-coding gene in a genome. Thus, RNA-seq can provide the 5′TSS, 5′UTR, exon–intron boundaries, 3′UTR, polyadenylation site, and alternative usage of any of these if applicable. A graphical example of a gene structure and what RNA-seq can identify is shown in Figure 1.4.

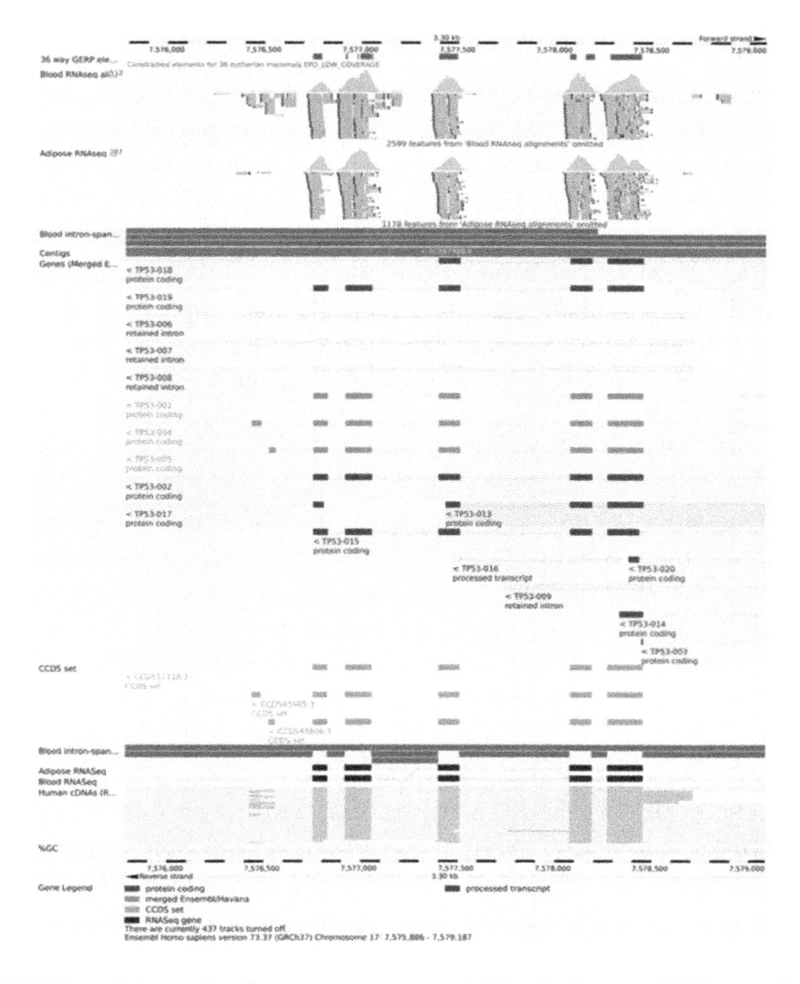

FIGURE 1.4 Schematic of gene structure model for the human TP53 gene from ENSEMBL genome browser showing RNA-seq reads from blood and adipose tissue as support for the models.

1.6.2 Novel Protein-Coding Genes

Early annotations of protein-coding genes relied on computational predictions based on genomic sequences. This was fine as long as genome data were available, the gene model elements fit common expected size and distance parameters, and there were transcriptomic data in the form of expressed sequence tag (EST) data sets or orthology data available to verify the predictions. However, it was easy to see that these criteria fit well only a very limited number of organisms under scientific investigation. Therefore, RNA-seq, with its high throughput, could verify many of the previous predictions, but also in places where no prediction existed, could identify novel protein-coding genes. It was especially useful in cases where no genome sequence was available, so a transcriptome of an organism could be built entirely from RNA-seq data. A recent example of this application has been in the sequencing of the black-chinned tilapia, an invasive fish of African origin with very scarce genomic resources [5]. Another example would be the oat (*Avena sativa* L.) transcriptome. Despite its delicious and nutritious character and economic importance, the allohexaploid genome has made it highly challenging to genetically map, sequence, and characterize. A recent RNA-seq study produced 134 M paired-end 100 nt reads and tripled the number of EST sequences available [6].

1.6.3 Quantifying and Comparing Gene Expression

Once the sequence and gene structure have been elucidated, it is logical that abundance values can be attributed to each gene as well as various features in their structures. As many studies would like to compare the abundance of RNA transcripts from healthy versus sick, nontreated versus treated, or time point 0 versus 1, it is logical that comparative studies are made. The range and types of comparative studies are virtually unlimited, so it is not productive to list them here. Instead, some relevant RNA-seq studies will be shown to illustrate the applications of RNA-seq to provide the reader with a flavor for what RNA-seq can provide.

In one of the earliest RNA-seq studies, transcripts from adult mouse brain, liver, and skeletal muscle were sequenced and compared [7]. More than 40 M single-end reads at 25 nt were sequenced on an Illumina platform and the authors found novel TSSs, alternative exons, and alternative 3′UTRs. The study demonstrated the shallowness of previous annotations of gene structure and thus highlighted how the breadth and depth of annotations provided by RNA-seq technology could change our view of gene structure. These results thus paved the way for subsequent RNA-seq studies.

Barely 2 years later, an RNA-seq study followed the expression of RNA transcripts from mouse skeletal muscle C2C12 cells during differentiation after 60 h, 5 days, or 7 days [8]. The technology improved so that >430 M paired-end reads at 75 nt were used to identify >3700 previously unannotated transcripts. TSSs were also shown to change in >300 genes during differentiation further demonstrating the extent of additional transcript knowledge RNA-seq could uncover in a relatively well-known cell culture system.

It is also possible to study RNA transcripts in whole animals. *Caenorhabditis elegans*, a free-living soil-dwelling nematode, was grown in 0.2 M ethanol or water from embryo stage to the final larva stage before becoming adults. The total RNA from whole animals were isolated and subjected to RNA-seq [9]. Over 30 M reads from water- or ethanol-treated animals were obtained. Ethanol exposure could be seen to increase RNA transcripts of detoxification enzyme genes and decrease transcripts involved in endoplasmic reticulum stress. Similar studies have also been performed on different model organisms with exposure to toxins that have ranged from carcinogens such as aflatoxin, and benzo(A)pyrene to the environmental-contaminant methylmercury.

In a recent evolutionary, model organism, and commercial application, the freshwater prawn (*Macrobrachium rosenbergii*) was subjected to RNA-seq analysis [10]. The polyA+-enriched RNA from total RNA of hepatopancrease, gill, and muscle yielded 86 M paired-end 75 nt reads. As this organism's genome had not been previously sequenced, the data were used to construct a transcriptome that consisted of >102,000 UniGenes of which 24% could be mapped to NCBI nr, Swissprot, KEGG, and COG databases.

1.6.4 Expression Quantitative Train Loci (eQTL)

RNA-seq studies have become so pervasive that they have been used to study quantitative traits. Traditionally, quantitative train loci studies in the form of genome-wide association studies have linked SNPs with a quantitative trait such as height, weight, cholesterol level, or risk to obtain type II diabetes. eQTL provides gene expression changes that can be correlated with known SNPs [11]. The basis for this correlation can be a local action, call *cis*-eQTL, for example, where an SNP is located on an enhancer region and changes the expression, or a distal action, called *trans*-eQTL, for example, where an SNP changes the structure of a transcription factor that no longer works on its target gene.

Thus, gene expression levels, as determined by RNA-seq, can provide a link with phenotype through its correlation with SNPs. An extension of this idea has been to correlate also gene-splicing sites and usage with SNPs. This approach, termed sQTL, suggests splicing as playing a significant role in regulating overall gene expression [12]. In addition to human disease research, this approach has been applied in traditional fields such as plant breeding where quantitative traits are important.

1.6.5 Single-Cell RNA-seq

Single-cell RNA-seq is a variation of RNA-seq where the source of total RNA for sequencing comes from a single cell. Typically, total RNA is not isolated, but rather cells are individually harvested from their source and reverse-transcribed. Methodology for library preparation is similar to RNA-seq: RNA is reverse-transcribed to cDNA, adaptors are ligated, barcodes for each cell are added, and ds cDNA amplified. Due to the low complexity of RNA species, single isolated cells or individual libraries are sometimes pooled prior to sequencing. In one example of this approach, a single mouse blastomere was collected and RNA-sequenced from its contents. The authors found 5000 genes expressed and >1700 novel alternative splice junctions, indicating both the robustness of the approach as well as the complexity of splicing in a single cell [13]. In another example of the approach, single cells from the nematode *C. elegans* at an early multicell developmental stage were isolated and libraries prepared from total RNAs. New transcription of genes could be monitored at each individual stage of development via profiling the transcripts of individual cells [14].

1.6.6 Fusion Genes

As read numbers and lengths increased, and paired-end sequencing became available, the ability to identify rare, but potentially important transcripts increased. Such is the case with fusion genes, which are transcripts generated from the fusion of two previously separate gene structures. Fusion partners can contribute 5′UTRs, coding regions, and 3′polyadenylation signals. Conditions for this event to occur happen during genomic rearrangement found in cancer tissues and cells. Cytogenetic derangements such as genomic amplifications, translocations, and deletions can bring together two independent gene structures. For example, 24 novel and three known fusion genes were detected in three breast cancer cell lines using paired-end sequencing of libraries sized 100 or 200 nt in length [15]. One of these fusion genes, *VAPB-IKZF3*, was found to be functional in cell growth assays. Recent RNA-seq studies have found fusion genes to be present in normal tissue, suggesting that fusion gene events might have normal biological function as well.

1.6.7 Gene Variations

As the amount of RNA-seq data accumulates, it is possible to mine the data for gene variation. Such an area is very active as data from large-scale projects and the published literature allow and even require data to be public. Mostly bioinformatic approaches by downloading publicly available data have been used to scan SNPs in transcriptomic data [16]. In this study, 89% of SNPs derived from RNA-seq data at a coverage of 10× were found to be true variants. SNP detection can also be obtained directly from original RNA-seq data. A group performed RNA-seq on muscle from *Longissimus thoraci* (Limousine cattle) muscle mRNAs [17]. They were able to identify >8000 high-quality SNPs from >30 M paired-end reads. A subset of these SNPs was used to genotype nine major cattle breeds used in France, demonstrating the utility of this approach.

One recent application of NGS has been to identify variations in the protein coding gene sequences from genomic DNA samples. Termed "exome-sequencing or exome-capture," this approach is technically not RNA-seq since it relies on sequencing fragmented genomic DNA that has been enriched for exons via hybridization to exonic sequences. This has been motivated by human disease studies, where variations, typically SNPs, need to be identified from a large cohort of individuals. Even today, sequencing cohorts of thousands of individuals is costly, so a shortcut is to sequence only the exonic sequences of an individual. As exons are overwhelmingly located in protein-coding genes, this has the advantage of finding variations that have direct effects on protein structure. It is one of the most popular applications of NGS and many commercially available kits have been developed for this purpose.

1.6.8 Long Noncoding RNAs

Another application of RNA-seq has been to find transcripts that are present, but do not code genes. Long noncoding RNAs (lncRNAs) were known before RNA-seq technologies were available. However, the extent of their existence and pervasiveness was not fully appreciated until RNA-seq methods were able to uncover the many different species of lncRNAs in living cells. lncRNAs are generally described as transcripts that fall outside of known noncoding RNAs such as tRNAs, ribosomal RNAs, and small RNAs, do not overlap a protein-coding exon, and are >200 nt in length [18]. lncRNAs can control transcription as enhancers (eRNA) epigenetically by binding and altering the function of histone proteins, as competitors to RNA-processing machinery [competitive endogenous RNA (ceRNA)], or as noise generated randomly. It can now be appreciated that lncRNAs may play a role in disease such as Alzheimer's disease [19].

1.6.9 Small Noncoding RNAs (miRNA-seq)

Finally, RNA-seq can be used to identify the sequence, structure, function, and abundance of small noncoding RNAs. The most well-known example of these being miRNAs (miRNA-seq), but other small noncoding RNAs such as small nuclear/nucleolar RNAs (snRNA), microRNA offset RNAs (moRNAs), and endogenous silencing RNAs (endo-siRNAs) can also be studied using miRNA-seq approaches. The methods used for miRNA-seq are similar to RNA-seq. The starting materials can be total RNA or size-selected/fractionated small RNAs. Most of the common sequencing platforms will sequence small RNAs once converted into ds cDNAs, such that much of the difference in the experimental protocols occur before sequencing. These will be described in detail in later chapters. Needless to say, there are many applications for characterizing these molecules not only in the studies of basic biochemistry, physiology, genetics, and evolutionary biology, but also in medicine as a diagnostic tool for cancer or in aging processes. A recent study of the nematode *Panagrellus redivivus* has presented the identification of >200 novel miRNAs and their precursor hairpin sequences while also providing gene structure models, annotation of the protein-coding genes, and the genomic sequences in a single publication [20].

1.6.10 Amplification Product Sequencing (Ampli-seq)

It is sometimes the case that whole transcriptomes do not need to be sequenced, but only a small number of genes. While one can always obtain a subset of genes of interest from a whole transcriptome sequence analysis, the effort, time, and resources required may be more than necessary. By using a panel of PCR primers consisting of 10–200 pairs, one can perform reverse transcription-PCR (RT-PCR) and instead of cloning each individual product and isolating plasmid DNA for Sanger sequencing, one can sequence the pool of PCR products to obtain the sequence. This has practical applications where the number of samples to be interrogated is large, and the number of genes is small.

1.7 CHOOSING AN RNA-SEQ PLATFORM

Now that the platforms have been described and some typical applications have been presented, it is natural to ask which platform should be chosen for a specific application. A simple solution would be to track down a PubMed reference based on the same or similar application and to choose based on published experience. It is of course always recommended to check the literature before embarking on a scientific study to see how past studies have dealt with the current problem. However, a weakness to blindly following past precedent is that NGS sequencing, in general, and RNA-seq, in particular, is rapidly changing both in how experiments are designed and how they are executed. Because of this rapid technology evolution, it is fair to say that there is no single right answer for a specific problem. Moreover, many RNA-seq projects have multiple aims. For example, one might want to identify new gene fusion transcripts in a sample, quantitate the abundance of already known genes, and identify any SNPs in known genes.

Therefore, it is more rationale to provide guidelines based on general study-design principles, so that the user can both plan the project with confidence in expected outcomes and also understand why some choices are made. Tradeoffs in depth of coverage and number of platforms used in a study may need to be made, and since laboratories have limited resources, it is inevitable that tradeoffs will need to be made.

1.7.1 Eight General Principles for Choosing an RNA-seq Platform and Mode of Sequencing

1.7.1.1 Accuracy: How Accurate Must the Sequencing Be?

If the goal is to detect SNPs or single-nucleotide editing events in the RNA species, then we must choose a platform that has a low error rate, in practice we should be able to distinguish between genuine SNPs and sequencing errors. With the human SNP frequency about 1/800, this corresponds to an accuracy rate of 99.9%. Only SOLID platform claims to have accuracy rates that exceed this level, and some platforms are far worse. However, we should keep in mind that we can compensate for low accuracy by having more reads. So 10 reads of the same piece of RNA with an accuracy of 99.9% can effectively provide an accuracy level of 99.99%.

If the goal is to identify known protein-coding genes and improve the annotation of the gene structure model as well as quantitate transcripts and perhaps discover new genes, then we need very little accuracy. In fact, programs to map reads to known gene models allow one or even two mismatches for a match. In effect, we are allowing for 98% accuracy if our reads are 50 nt long and allow one mismatch. At this level, most of the common platforms can be used: SOLID, Illumina, 454, Ion Torrent.

1.7.1.2 Reads: How Many Do I Need?

It is good practice to calculate the coverage statistics in our RNA-seq study. As a rough calculation, the human genome has 3000 M nt of which approximately 1/30 is used for protein-coding genes. This means the RNA to be sequenced is represented in about 100 M nt. If we are using single read 100 nt (or paired-end read 50 nt), then 1 M reads gives 100 M nt of sequence data which equals 1× coverage. A total of 30 M reads, which is a typical read output from the common platforms, would provide 30× coverage. So with 30 M reads, we can expect to have a huge amount of reads for abundantly expressed genes, good coverage for most genes, and may miss a few low expressed or rarely expressed genes. To calculate the probability that a read will map to a specific gene, we can assume an average gene size of 4000 nt (100 M nt divided by 25,000 genes). At 30 M reads equivalent to 30× coverage, at single read 100 nt (or paired-end read 50 nt) length, we can expect a single read to map to the average expressed and length gene 4000 nt× 30 coverage/100 nt 1200 times. Thus, if the gene is expressed at a level of 1/1200 compared to the average gene, then we have a 50:50 probability to have a read map to it. In practice, 30 M reads is quite reasonable to capture most, but probably not all of the genes expressed in a sample. As most of the platforms can produce up to 30 M reads, this is usually not a limitation. Where better coverage is needed, and data for alternative exon usage and other gene model details or rare events are needed, then the platforms that more easily produce a large number of reads are preferred. A recently developed method called "capture-seq" has been used to enrich the RNAs in a small number of loci of the human genome. The method essentially uses a printed Nimblegen microarray to capture RNAs from a limited number of loci [21]. In the example, the authors captured approximately 50 loci including protein-coding genes and long noncoding RNAs. With the capture strategy, they were able to effectively obtain >4600-fold coverage of their loci and were able to discover unannotated exons and splicing patterns, for even well-studied genes. The simple conclusion is that you may never have enough coverage to obtain every single possible transcript from a locus.

Another way of looking at the problem is to consider how many reads are necessary to confirm the existence of a transcript. There is no consensus on this matter and the literature is full of examples where one read is sufficient to claim the existence of a molecule, and in contrast, literature where <10 reads is not enough. Much of this depends upon the context of the study, the journal or database criteria, and the overall aims of the study.

1.7.1.3 Length: How Long Must the Reads Be?

For simply mapping reads to known gene models in an organism, even 14 nt is sufficient. However, because some reads may map to >1 location, longer reads are needed. At 50 nt, a small percentage of reads will still map to >1 location, but the number is typically quite small (<0.01%), so practically this read length will allow you to make differential expression studies and better define gene models. There are, however, many cases where longer reads are necessary such as annotating novel genes in a species for which no other sequence data (e.g., genomic, EST, or long cDNAs) are available. Having longer sequences, rather than trying to predict gene models based on mapping discontinuous 50 nt reads, is a distinct advantage. Roche 454 has an established track record for these types of applications. Pacific Biosciences, especially the newer generation instruments and kits, are able to produce long reads at up to 10,000 nt or more routinely.

1.7.1.4 SR or PE: Single Read or Paired End?

If there were no biases in any of the steps of library preparation (fragmentation of RNA, ligation of adaptors, orientation of strands), and cDNA synthesis would produce completely random fragments representative of the RNA sample, then we would obtain the same sequence information from SR as we would from PE. However, there are biases in these library preparation steps. One way of increasing the randomization of fragments to be sequenced is by sequencing both ends of a library clone. This serves a dual purpose in that PE sequences from short fragments can overlap providing additional confirmation of a sequence. Most data analysis programs are now capable of handling both SR and PR data, so there is not even a hindrance in downstream analysis. Unfortunately, not all platforms allow for sequencing at both ends, so essentially, if it is available, it is a good idea to use paired-end.

1.7.1.5 RNA or DNA: Am I Sequencing RNA or DNA?

As mentioned earlier, most platforms sequence double-stranded cDNA derived from reverse transcription and PCR amplification of RNA molecules in a sample. There are instances in RNA-seq where RNA would preferably be sequenced directly such as in projects where modification to RNA structures is important such as mRNA capping.

1.7.1.6 Material: How Much Sample Material Do I Have?

These days, as it is possible to sequence total RNA from a single cell, one wonders whether there is a lower limit to the sample material needed. Sequencing platforms that use amplified, double-stranded cDNA have essentially no lower limit on material. However, this does not mean one should provide the sequence platform with the minimum amount. Increasing material also should increase the representation of RNA species in a sample. Most of the sequencing-by-synthesis platforms have now specialized kits for making libraries from nanogram amounts of total RNA. Single-molecule platforms by definition need only a single molecule for sequencing. Therefore, this does not appear to be a limitation for the different platforms.

1.7.1.7 Costs: How Much Can I Spend?

As the costs of sequencing have come down dramatically in the past 10 years, cost should not be a consideration. However, the reality is that the requirements to publish and the standards for quality continue to go up as well, so there is always the issue of costs. Sending RNA-seq libraries to commercial, national, or local core NGS facilities is a good way of mitigating costs. If one is well funded, buying a personal laboratory sequencer is nowadays feasible. In fact, both Illumina with MiSeq and Ion Torrent with the Personal Genome Machine and Ion Proton have produced personal laboratory sequencers that are now affordable for even not so well-funded laboratories. The lower end of the price scale has probably not been reached yet, so one can look forward to having even more choices in sequencing platforms that are not economically motivated. Indeed, the high degree of solicitation for samples by commercial and nonprofit core facilities would indicate that pricing pressures continue to go downward.

1.7.1.8 Time: When Does the Work Need to Be Completed?

There is an old adage in work life that the task needed to have been completed "yesterday." Genomics is a fast-moving field and ideally samples are prepared, libraries are constructed, and sequencing is performed without any queues or delays. In reality, many platforms (Illumina, SOLID, 454) have queues, not because the machine is running, but because not enough libraries sufficient to fill a flow cell for a single run have been constructed and submitted for sequencing. Suffice to say that, in practice, work queues may arise not from the instrument, but because of preparation work in constructing libraries and collecting a sufficient number of libraries to start the instrument run. On the other end of the work flow, once sequence data are generated, the work is just beginning and data analysis can commence. The data analysis stage may then take days, months, or years in the case of big projects that make the sequencing instrument run time appear relatively short in comparison.

1.7.2 Summary

In conclusion, one can see the vast number of options available for performing RNA-seq experiments. Each platform has its own unique properties that differentiate it from others. A list showing the major RNA-seq platforms, their sequence and detection chemistries, and web links can be found in Table 1.1. If fortunate, one has multiple platforms to choose from. Indeed, some studies take advantage of the best properties of each platform so that different ones are justified for different purposes. For example, Illumina reads might be used for coverage, SOLID for accuracy, and Roche 454 or Pacific Biosciences for length. One can easily imagine a future where multiple platform usage is typical for a specific project. The factors in choosing a platform are multidimensional, but it is not impossible to determine the most appropriate platform for a particular application. Utilizing the information presented here, along with updated specifications for the instruments and current pricing, it should be possible to make an informed decision on the appropriate platform to use and their mode of usage for RNA-seq experiments.

REFERENCES

1. Nagalakshmi U., Wang Z., Waern K. et al. The transcriptional landscape of the yeast genome defined by RNA sequencing. *Science* 320(5881):1344–1349, 2008.
2. Sultan M., Schulz M.H., Richard H. et al. A global view of gene activity and alternative splicing by deep sequencing of the human transcriptome. *Science* 321(5891):956–960, 2008.
3. Wilhelm B.T., Marguerat S., Watt S. et al. Dynamic repertoire of a eukaryotic transcriptome surveyed at single-nucleotide resolution. *Nature* 453(7199):1239–1243, 2008.
4. Wang Z., Gerstein M., and Snyder M. RNA-Seq: A revolutionary tool for transcriptomics. *Nature Reviews in Genetics* 10(1):57–63, 2009.
5. Avarre J.C., Dugué R., Alonso P. et al. Analysis of the black-chinned tilapia *Sarotherodon melanotheron heudelotii* reproducing under a wide range of

salinities: From RNA-seq to candidate genes. *Molecular Ecology Resources* 14(1):139–149, 2014.

6. Gutierrez-Gonzalez J.J., Tu Z.J., and Garvin D.F. Analysis and annotation of the hexaploid oat seed transcriptome. *BMC Genomics* 14:471, 2013.
7. Mortazavi A., Williams B.A., McCue K. et al. Mapping and quantifying mammalian transcriptomes by RNA-seq. *Nature Methods* 5(7):621–628, 2008.
8. Trapnell C., Williams B.A., Pertea G. et al. Transcript assembly and quantification by RNA-seq reveals unannotated transcripts and isoform switching during cell differentiation. *Nature Biotechnology* 28(5):511–515, 2010.
9. Peltonen J., Aarnio V., Heikkinen L. et al. Chronic ethanol exposure increases cytochrome P-450 and decreases activated in blocked unfolded protein response gene family transcripts in *Caenorhabditis elegans. Journal of Biochemical Molecular Toxicology* 27(3):219–228, 2013.
10. Mohd-Shamsudin M.I., Kang Y., Lili Z. et al. In-depth transcriptomic analysis on giant freshwater prawns. *PLoS ONE* 8(5):e60839, 2013.
11. Majewski J. and Pastinen T. The study of eQTL variations by RNA-seq: From SNPs to phenotypes. *Trends in Genetics* 27(2):72–79, 2011.
12. Lalonde E., Ha K.C., Wang Z. et al. RNA sequencing reveals the role of splicing polymorphisms in regulating human gene expression. *Genome Research* 21(4):545–554, 2011.
13. Tang F., Barbacioru C., Wang Y. et al. mRNA-seq whole-transcriptome analysis of a single cell. *Nature Methods* 6:377–382, 2009.
14. Hashimshony T., Wagner F., Sher N. et al. CEL-Seq: Single-cell RNA-seq by multiplexed linear amplification. *Cell Reports* 2(3):666–673, 2012.
15. Edgren H., Murumagi A., Kangaspeska S. et al. Identification of fusion genes in breast cancer by paired-end RNA-sequencing. *Genome Biology* 12(1):R6, 2011.
16. Quinn E.M., Cormican P., Kenny E.M. et al. Development of strategies for SNP detection in RNA-seq data: Application to lymphoblastoid cell lines and evaluation using 1000 Genomes data. *PLoS ONE* 8(3):e58815, 2013.
17. Djari A., Esquerré D., Weiss B. et al. Gene-based single nucleotide polymorphism discovery in bovine muscle using next-generation transcriptomic sequencing. *BMC Genomics* 14(1):307, 2013.
18. Ilott N.E. and Ponting C.P. Predicting long non-coding RNAs using RNA sequencing. *Methods* 63(1):50–59, 2013.
19. Faghihi M.A., Modarresi F., Khalil A.M. et al. Expression of a noncoding RNA is elevated in Alzheimer's disease and drives rapid feed-forward regulation of beta-secretase. *Nature Medicine* 14(7):723–730, 2008.
20. Srinivasan J., Dillman A.R., Macchietto M.G. et al. The draft genome and transcriptome of *Panagrellus redivivus* are shaped by the harsh demands of a free-living lifestyle. *Genetics* 193(4):1279–1295, 2013.
21. Mercer T.R., Gerhardt D.J., Dinger M.E. et al. Targeted RNA sequencing reveals the deep complexity of the human transcriptome. *Nature Biotechnology* 30(1):99–104, 2011.

CHAPTER 2

Introduction to RNA-seq Data Analysis

2.1 INTRODUCTION

Once the millions of reads are obtained from an RNA-seq experiment, the data analysis begins. As described in Chapter 1, RNA-seq is a powerful technology with many applications, ranging from gene and splice variant discovery to differential expression analysis and detection of fusion genes, variants, and RNA editing. Consequently, there are many data analysis paths available and it is not possible to cover them in a single analysis workflow scheme. The overview in Figure 2.1 covers the major steps in most of the routine analyses, where different paths are followed depending on whether a reference genome or transcriptome is available or not. The data analysis steps are performed by separate programs, which may need specific data formats and external files. As RNA-seq data analysis is an active field of research producing new approaches and tools at a rapid pace, many alternative programs exist for each analysis step. Keeping track of the options available and selecting the most suitable program can be a challenge, but luckily many thorough tool comparisons have been published, and we refer to these evaluation articles in the following chapters.

So how do we start RNA-seq data analysis? This depends on what kind of analysis you would like to perform and what is your background. If you are not comfortable working on command line and using Unix and R, you might like to opt for a user-friendly graphical user interface such as

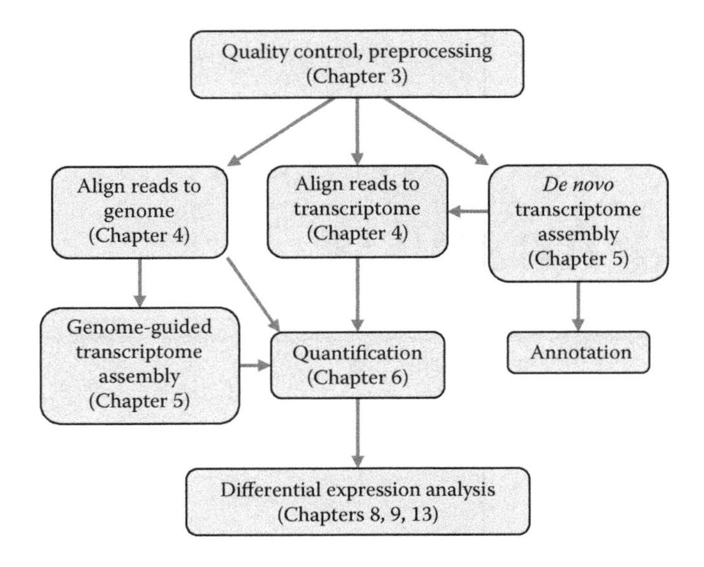

FIGURE 2.1 Possible paths in RNA-seq data analysis. In the beginning, the quality of reads is checked and if necessary, reads are preprocessed to remove low-quality data and artifacts. The read's origin is then identified by aligning them to a reference genome if available. Novel genes and transcripts are detected using genome-guided transcriptome assembly, and gene and transcript expression are quantified. Alternatively, gene and transcript discovery can be skipped and expression quantified only for known genes and transcripts. If the reference genome is not available, reads can be aligned and quantified using a reference transcriptome instead. If a transcriptome is not available, it can be produced from reads using *de novo* transcriptome assembly. When abundance estimates are obtained using one of these paths, expression differences between sample groups can be analyzed using statistical testing. The details of each step can be found in the chapters indicated in parentheses.

Galaxy (galaxyproject.org/) [1] or Chipster (chipster.csc.fi) [2]. These are integrated and flexible tools which can take one from raw RNA-seq reads to experimental results, and they are very handy from a practical point of view. An example of the Chipster interface is shown in Figure 2.2.

However, many users want to have a complete control over their data analysis and maximum flexibility to change parameters, use all available options, and import and export data to different tools of which some are standard and some are not. For these users, it will be necessary to become familiar with a command line environment. A good knowledge of Unix commands is helpful, and there are excellent resources available on the Web, such as www.ks.uiuc.edu/Training/Tutorials/Reference/unixprimer.

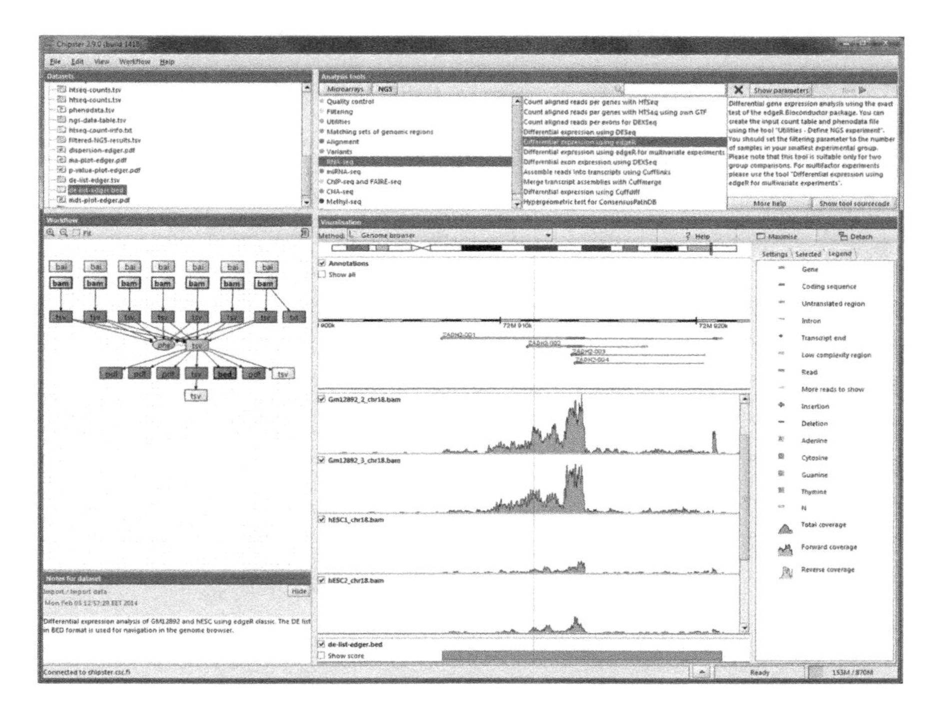

FIGURE 2.2 The open source Chipster software provides a comprehensive collection of analysis tools for RNA-seq data via an intuitive graphical user interface. The workflow panel (bottom left) shows the relationships of result files. This screenshot shows a differential expression analysis of GM12892 and hESC cells, which are used as an example throughout the book. Gene-level expression level estimates were obtained using genome-aligned reads and the HTSeq tool, and differential expression was analyzed using the edgeR Bioconductor package. Differentially expressed genes were further filtered by fold change and visualized in the built-in genome browser.

html. Many RNA-seq data analysis programs are written in Java, Perl, C++, or Python. Simply running these programs does not require programming knowledge, although it is of course helpful. Indeed, most tools these days have very good help manuals for installing and running the programs. For the examples in this book and the tools currently in use, some familiarity with R is important. For those who are not familiar with R, we suggest them to take advantage of some freely available web resources such as those at http://www.ats.ucla.edu/stat/r/. To those who are interested only in learning about the methodology in RNA-seq without performing hands-on experiments and analysis, then all that is required is a flexible and open mind.

2.2 DIFFERENTIAL EXPRESSION ANALYSIS WORKFLOW

In the previous section, we provided a general scheme for data analysis. In this section, we provide short descriptions of the major steps for differential expression analysis, which is a very common task in RNA-seq. The example workflow assumes that a reference genome is available. For each step, we describe the analysis goals, some typical options, inputs and outputs, and point out where a full chapter describing that step can be found. We hope to provide an overview of the entire data analysis process so that the user can see how individual steps are related to one another. Figure 2.3 provides an overall view of the data-handling steps. Separate programs exist for each step, although some tools can combine a few of them.

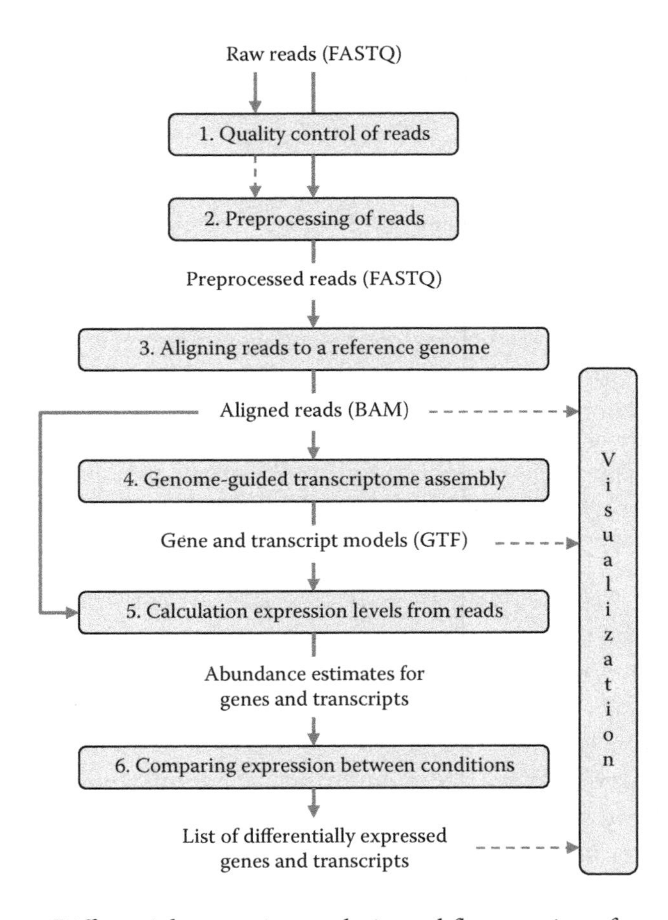

FIGURE 2.3 Differential expression analysis workflow consists of several, interrelated steps. The typical output file formats are indicated in parentheses.

2.2.1 Step 1: Quality Control of Reads

The analysis starts with raw sequence reads, typically in FASTQ format, although other formats can sometimes be used. When other formats are not supported by the program, the reads must be reformatted to FASTQ. In the first step, a general quality control analysis is performed. This looks at the overall quality of the millions of reads. Reads are scanned for low-confidence bases, biased nucleotide composition, adapters, duplicates, etc. as described in Chapter 3. The output of this step is basic statistics such as the number of reads and quality information which guides the preprocessing decisions in the next step.

2.2.2 Step 2: Preprocessing of Reads

The goal of preprocessing is to remove low-quality bases and artifacts such as adapter or library construction sequences from individual reads. Experimental artifacts can also be removed. For example, poly A tails can be removed since they interfere with analysis steps later. Another source of artifacts is microbiome which is present among many living organisms. Removing *Escherichia coli* sequences from human RNA tissue samples may help in later downstream steps. Reads may be trimmed also because of their size. For example, mature microRNA sequences are 21–22 nt in length, while the read length could be 50 nt. Preprocessing by trimming and filtering is covered in Chapter 3, whereas error correction is covered in Chapter 5 in the context of transcriptome assembly. After preprocessing, the data are now in a cleaned and polished form that can undergo the next data analysis steps.

2.2.3 Step 3: Aligning Reads to a Reference Genome

The goal of this step is to find the point of origin for each and every read. If a reference genome is not available yet, reads can be mapped to a transcriptome (which can be created *de novo* from the reads if necessary as described in Chapter 5). When a read is mapped to a reference, a sequence alignment is created. It is necessary to have a reference sequence as one of the input files in this step, in addition to the file of the preprocessed reads. Mapping is computationally intensive because there are millions of reads to map, genomes are large, and spliced reads have to be mapped noncontiguously. Therefore, the genome sequence is often transformed and compressed into an index to speed up mapping. The most common one in use is the Burrows–Wheeler transform. The output from this step is an alignment file, which lists the mapped reads and their mapping positions in the reference. In addition to downstream analysis, aligned reads can be visualized in genomic context using one of the genome viewers. Mapping and genomic visualization are covered in Chapter 4, which also discusses some utilities for manipulating the alignment files.

2.2.4 Step 4: Genome-Guided Transcriptome Assembly

If reads were aligned to a genome, the alignments can be used for discovering new genes and splice variants. Genes are large compared to sequence reads. For example, mature mRNA from mammalian species is typically 1.5 kb compared to the 100–250 nt of RNA-seq reads. It therefore is not normally possible to know the exact structure (transcription start site, exon–intron organization, poly A location) of a transcript from a single read. Longer read platforms such as PacBio Systems can actually sequence through an entire transcript, but shorter read data still dominate analysis at the moment. Most exons are <200 bp, so the use and order of alternative exons need to be reconstructed from mapping to the genome, and then linking alignments from one region to another. This is again simple on the surface, but computationally demanding, and therefore has required some skillful workarounds which are detailed in the Chapter 5 on transcriptome assembly. The output of this step is gene and transcript models. Assembled transcripts from different samples are merged and combined with reference annotation in order to produce more complete gene models, which can then be used for expression quantification in the next step.

2.2.5 Step 5: Calculating Expression Levels

A key data table that is generated by the analysis is the number of reads per gene and transcript. In this step, a single read is associated with a gene based on its mapping location. Expression of novel genes and transcripts can be quantified using the gene models obtained in the previous step. When working with data from a well-annotated organism such as humans, reference annotation can be used instead, thereby limiting the quantification to only known genes and transcripts. Abundance estimates can be reported in raw read counts or in normalized units such as RPKM (reads per thousand nucleotides in transcript per million reads) or FPKM (fragments per thousand nucleotides per million mapped reads). Information on the number of reads mapping to genes and different genomic feature types enables important quality control metrics too, as described in Chapter 6. At this step, the data become simply a table of genes and their read counts or RPKM/FPKM values.

2.2.6 Step 6: Comparing Gene Expression between Conditions

Once we have the abundance information, we can compare the values between groups of samples using statistical testing. Normalization is necessary because of the possible differences in read numbers and transcriptome composition, and many statistical tools include built-in normalization methods. Statistical methods are covered in Chapters 8 and 9.

2.2.7 Step 7: Visualization of Data in Genomic Context

It is important to visualize reads and results in a genomic context during the different stages of analysis in order to gain insights into gene and transcript structure and to obtain a sense of abundance. Many genome browsers are available and these can be used with one's own data or from preloaded data. One example is the Integrative Genomics Viewer (IGV) which allows one to view the RNA-seq as well as other genomic data [3]. In Figure 2.4, we loaded read alignment files for two samples, one from control and one from ethanol-treated animals. Notice the larger number of mapped reads from the control sample. Genome browsers are covered in Chapter 4.

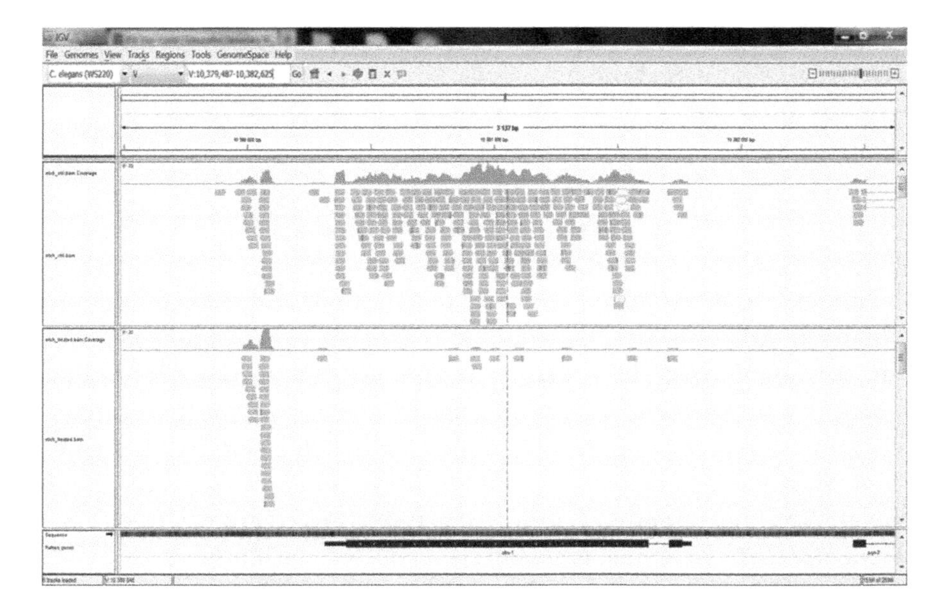

FIGURE 2.4 Integrative Genomics Viewer (IGV) window showing RNA-seq reads from the *C. elegans abu-1* gene. The top panel shows reads from control and the bottom panel shows reads from ethanol-treated animals.

2.3 DOWNSTREAM ANALYSIS

2.3.1 Gene Annotation

A typical output of a transcriptome study, for which the genome has been sequenced, is a list of expressed known genes. Reads may provide more information regarding the gene structure such as alternative transcription start sites and new exons. New genes may also be discovered. Output for novel genes is typically a provision gene identifier and a sequence. The user must then compare the sequence to known genes with tools like BLAST in order to identify the function of the gene. In addition to protein-coding genes, other types of novel transcripts can also be identified, such as long noncoding RNAs.

In cases where the RNA-seq data represent the first description of genes from an organism, an automated pipeline is set up to annotate the genes. Annotation is performed on computationally predicted genes built up from the reads that are aligned to create long transcripts. Initially, structural RNA molecules are annotated (tRNAs, ribosomal RNAs, snoRNAs, etc.) and removed, and then protein coding genes are aligned to databases of known genes and their function inferred from sequence similarity.

2.3.2 Gene Set Enrichment Analysis

The output of differential expression analysis is typically a list of genes and their expression level differences between two or more groups. The user can apply different cut-off thresholds, for example, 2-fold difference or a p-value <0.01, to shorten the list into a workable number of genes. Even with strict criteria, the user may be left with hundreds of genes, so making sense of the data is difficult. Gene set enrichment analysis provides a means by which the genes in a data set can be grouped based on their annotation and tested for overrepresentation in a group compared to a background such as all genes. For example, a gene list might contain 20 transcription factors out of a list of 200 genes. Is this significant when the genome contains 800 transcription factors out of 22,000 genes? The most common annotation used for grouping genes is the Gene Ontology [4], which is a controlled vocabulary. It is hierarchical, so one can annotate a gene at different levels of detail, and a gene can have multiple annotations. For example, a transcription factor can also be a nuclear factor, a receptor, and a DNA-binding protein. Gene set enrichment analysis provides the list of over-represented molecular functions, biological processes, and cell locations that can then be used to test the hypothesis of whether genes in a certain biochemical or cellular pathway are dysregulated.

2.4 AUTOMATED WORKFLOWS AND PIPELINES

It is often desirable to automate the handling of multiple data analysis steps. For routine RNA-seq analysis, it is possible to script a reusable analysis pipeline with defined stages, inputs, outputs, and parameters for each step. It is more challenging to get all the analysis steps within a single graphical user interface, because of the complexity of analysis and the large number of tools. However, great strides are being made in this direction, and it is already possible to perform the most common and routine RNA-seq analysis within a single software tool. These tools use established programs that can be brought together into a workflow in a single graphical user interface. Two examples that allow the user to build their own next-generation sequencing workflows include Galaxy (galaxyproject.org/) [1] and Chipster (chipster.csc.fi) [2] as shown earlier. In addition to these tools, commercial tools are also available. These tools provide a handy graphical user interface and ease of use, but at a cost.

2.5 HARDWARE REQUIREMENTS

One should be aware that RNA-seq produces a huge amount of data. Just one sample can generate 60 M reads of 100 bases (0.6 gigabases of sequence) which require several gigabytes (GB) of storage. Hardware requirements depend on the size and type of the experiment. If no *de novo* transcriptome assembly is performed, a small-scale study can be conducted on a desktop workstation with 4 GB of random-access memory (RAM), 200 GB of hard disk space, and 2.5 GHz clock speed. In this situation, mapping reads to a genome will take at least overnight. As more complex experiment designs containing more conditions and replicates are used, week-long run times for mapping are common. At this level, one should consider a minimum of 16 GB of RAM, 2 TB external hard drive or 48 TB server, and 3.6 GHz clock speed if one wants to run it on one's own workstation.

When a core or service facility is responsible for the analysis, standard workstations are simply not sufficient to handle the throughput needed in a reasonable amount of time. In the case of an RNA-seq data analysis core facility, a Linux cluster of >200 processors and 1 petabyte of storage is recommended. However, one should be aware of another solution, cloud services, that has essentially unlimited amount of storage and computing power. One of the best features of the cloud is the ability to rent data storage space and obtain processing power on an as-needed basis.

One should also be aware of data transfer rates from wherever the raw RNA-seq data is residing to the analysis environment. Even with the best optic fiber infrastructure of 1 Gbit/s (125 MB/s) bandwidth, raw data transfer may last hours or days due to congestion, writing to hard disk, or limits placed by system administrators. From a practical standpoint, typical RNA-seq users do not have much control over the bandwidth available, but they can budget in and calculate the time needed for simple data transfer. In some situations, it is more feasible and faster to ask the data generation facility to mail hard disks containing the data to the user.

2.6 FOLLOWING THE EXAMPLES IN THE BOOK

This book contains a lot of examples on how to use different analysis tools. We recommend that one performs the examples oneself using the same data sets. The files are available for download at the book's website at http://rnaseq-book.blogspot.fi/. In order to enable both bioinformaticians and nonprogramming wet-laboratory scientists to follow the examples, we provide two sets of instructions for the same tasks. One set uses command line tools and R, whereas the other set uses the Chipster software which has a graphical user interface. All software used in the examples are open source and freely available.

2.6.1 Using Command Line Tools and R

We suggest that you install a Linux distribution such as Ubuntu, as most of the command line analysis tools that we demonstrate in this book run on the Linux operating system. If you are a Windows user and do not want to switch to Linux completely, you can create a disk partition so that you can run both Linux and Windows on your computer. For details on how to download Ubuntu, see http://www.ubuntu.com/download/desktop.

The examples cover a large number of analysis tools and it is beyond the scope of this book to include installation instructions for all of them, but detailed instructions are available on the tool websites. An alternative for downloading and installing each tool individually is to use the Chipster virtual machine, which is based on Ubuntu and contains most of the analysis tools and reference data sets. You can use the tools via a graphical user interface as described below, but you can equally well log in to the virtual machine and use them on command line. Instructions for setting up the virtual machine are given in the next section.

It is a good practice to take notes of the code that you have written, because it allows you to reproduce the analysis steps later. In addition to simple text editors like Notepad in Windows and nano in Linux, there are also some specialized code editors, such as Notepad++ and RStudio for R on Windows, and emacs (with some extra add-ons) on Linux. These specialized tools are strongly recommended, because they make code editing easier by coloring parts of the code differently, so, for example, it is easier to see commands and arguments.

Whenever you encounter new code or commands, you might want to get more information about the available options and inner workings. For R, you can access the built-in help with the command "?" followed by the name of command you are searching help for. For example, "?lm" would open the help page for the command lm, which fits linear models to data. For Unix commands, you can access a manual page with the command man. For example, "man less" opens the manual page for the command less. For command line analysis software, help is available at the software's home page. There are also active forums such as SEQanswers (http://seqanswers.com/) and Biostar (http://www.biostars.org/) where you can post data analysis questions.

2.6.2 Using the Chipster Software

If you would like to try the analysis examples in this book but you do not feel comfortable working with command line tools and R/Bioconductor, you can perform the same analysis tasks using the graphical Chipster software. Chipster is open source and is freely available. It provides a comprehensive collection of data analysis tools for different next-generation sequencing applications, including RNA-seq. The tools cover all the steps from the quality control to statistical testing and pathway analysis. You can start your analysis from any point, importing either raw reads (FASTQ), alignments (BAM), or a count table.

Technically Chipster is a Java-based client-server system, which is available as a virtual machine image at http://chipster.sourceforge.net/downloads.shtml. The virtual machine contains all the analysis tools and reference data sets, so it is ready to use (but relatively large). You need a virtualization software such as VMware or Virtual Box to run the Chipster virtual machine in Windows, MacOS X, or Linux. If you have not used virtual machines before, we recommend that you get a local system administrator to help you with the installation. If a local installation is not possible, you can use the Chipster server in Finland (http://chipster.

csc.fi/), although this is not optimal because of the data transfer times. If you just want to look at the ready-made analysis sessions or use the genome browser, you can log in with the username guest. Free evaluation accounts are also available.

Some general instructions for using Chipster can be found below, while instructions for the individual analysis steps are embedded in the relevant chapters. A screenshot of Chipster user interface is shown in Figure 2.1.

- Import data by selecting "Import files." The files will appear in both the Data sets view (top left) and the Workflow view (bottom left).
- Analysis tools are grouped in categories (top right). Every tool has a little help text, and you can get to the tool manual by clicking on the "More help" button. You can also see the source code of a tool by selecting the "Show tool source code" button. Tools can be run with default parameters, but it is recommended that you check if they are suitable for your data.
- In order to run an analysis, select the file, the tool category, and the tool. After checking and possibly changing the parameters, click "Run." You can monitor the status of the run by clicking on the small triangle in the bottom panel.
- When an analysis run has completed, the result files appear in the Data sets view and in the Workflow view. You can visualize the results by selecting a file and a suitable visualization method from the Visualization panel (bottom right).
- Save the analysis session by selecting "File/Save session." This will pack all the files, their relationships, and metadata (information on what tool and what parameters were used for creating a particular result file) to a zip file. You can also save a workflow, which allows you to run the same analysis steps with just one click on a different data set.
- Chipster keeps a "lab book" of what you have done. Clicking on the small paper icon in the Workflow panel produces a textual report listing all the steps (tools and their parameter settings) that led to a particular file.
- If you need any help with a particular functionality like the built-in genome browser, consult the manual (http://chipster.csc.fi/manual/) or send your question to the Chipster mailing list.

2.6.3 Example Data Sets

The examples use ENCODE data from GM12878 and H1-hESC cell lines. GM12878 is a lymphoblastoid cell line produced from the blood of a female donor by EBV transformation, and H1-hESC are human embryonic stem cells. The data were produced at Caltech and consist of 75 base paired-end reads with insert length 200. It is from an earlier Illumina platform, so the base quality encoding is phred64.

There are three GM12878 samples and four H1-hESC samples. Reads from the GM12878 replicate 2 are used in Chapters 3, 4, and 6. Chapter 5 uses reads from the H1-hESC replicate 1, concentrating on the reads which map to chromosome 18. Chapters 7–10 use reads that map to chromosome 18 from all the samples.

The files can be found at http://hgdownload.cse.ucsc.edu/goldenPath/hg19/encodeDCC/wgEncodeCaltechRnaSeq/.

The following FASTQ files were selected:

wgEncodeCaltechRnaSeqGm12892R2x75Il200FastqRd1Rep2V2.fastq.gz
wgEncodeCaltechRnaSeqGm12892R2x75Il200FastqRd2Rep2V2.fastq.gz

The following BAM files were selected:

wgEncodeCaltechRnaSeqGm12892R2x75Il200AlignsRep1V2.bam
wgEncodeCaltechRnaSeqGm12892R2x75Il200AlignsRep2V2.bam
wgEncodeCaltechRnaSeqGm12892R2x75Il200AlignsRep3V2.bam
wgEncodeCaltechRnaSeqH1hescR2x75Il200AlignsRep1V2.bam
wgEncodeCaltechRnaSeqH1hescR2x75Il200AlignsRep2V2.bam
wgEncodeCaltechRnaSeqH1hescR2x75Il200AlignsRep3V2.bam
wgEncodeCaltechRnaSeqH1hescR2x75Il200AlignsRep4V2.bam

In a couple of examples, we use a set of RNA-seq data from a study of primary cultures derived from parathyroid tumors. The raw data (as well as estimated expression values) are available from Gene Expression Omnibus (GEO accession GSE37211), but here we use an R/BioConductor package, *parathyroid*, which was developed by Michael Love and which contains an "analysis-ready" version of the data set, allowing straightforward loading of both the expression data and associated metadata into an R session. This data set contains RNA-seq measurements of mRNA levels in tumors cultured from four different patients. For each patient, two cultures were each treated with a different chemical (diarylpropionitrile or "DPN" and 4-hydroxytamoxifen or "OHT," respectively), while one culture was kept as a control. In addition, each culture was sampled at two time points, so that there are six measurements per patient (except in one case, where the library preparation failed and no usable sequence was obtained).

2.7 SUMMARY

RNA-seq is a powerful technology with many applications, and consequently there are many data analysis paths available. Even the most routine analyses require a large number of discrete steps that are interrelated. While seemingly complicated at first, one eventually can see the logic behind the various steps and how they relate to each other. A typical shorthand for the RNA-seq data analysis steps is mapping, transcript construction, and expression quantification [5]. We have endeavored to break this down into even smaller steps so that the reader can see both the generated data output and how they were derived. What we hope to demonstrate in the following chapters are the background, theory, and practical execution of each step involved in RNA-seq analysis. As RNA-seq data analysis is an active field of research producing new approaches and tools all the time, we recommend that you follow actively the literature and discussion forums such as SEQ answers and Biostar.

REFERENCES

1. Goecks J., Nekrutenko A., Taylor J. et al. Galaxy: A comprehensive approach for supporting accessible, reproducible, and transparent computational research in the life sciences. *Genome Biology* 11(8):R86, 2010.
2. Kallio M.A., Tuimala J.T., Hupponen T. et al. Chipster: User-friendly analysis software for microarray and other high-throughput data. *BMC Genomics* 12:507, 2011.
3. Thorvaldsdóttir H., Robinson J.T., and Mesirov J.P. Integrative Genomics Viewer (IGV): High-performance genomics data visualization and exploration. *Briefings in Bioinformatics* 14(2):178–192, 2013.
4. Ashburner M., Ball C.A. et al. The Gene Ontology Consortium. Gene ontology: Tool for the unification of biology. *Nature Genetics* 25(1):25–29, 2000.
5. Garber M., Grabherr M.G., Guttman M. et al. Computational methods for transcriptome annotation and quantification using RNA-seq. *Nature Methods* 8(6):469–477, 2011.

CHAPTER 3

Quality Control and Preprocessing

3.1 INTRODUCTION

Quality problems typically originate either in the sequencing itself or in the preceding library preparation. They include low-confidence bases, sequence-specific bias, 3′/5′ positional bias, polymerase chain reaction (PCR) artifacts, untrimmed adapters, and sequence contamination. These problems can seriously affect mapping to reference, assembly, and expression estimates, but luckily many of them can be corrected for by filtering, trimming, error correction, or bias correction. Some problems cannot be corrected for, but you should at least be aware of them when interpreting results.

This chapter covers the quality checking of raw reads, that is, FASTQ files [1]. Once reads have been aligned to a reference genome, additional quality metrics can be investigated based on the location information as discussed in Chapter 6. These include coverage uniformity along transcripts, saturation of sequencing depth, ribosomal RNA content, and read distribution between exons, introns, and intergenic regions. Finally, once aligned reads have been counted per genes, sample relations and batch effects can be visualized with heatmaps and PCA plots. This experimental-level quality control is discussed in conjunction with statistical testing in Chapter 8.

In addition to quality checking, this chapter also covers read trimming and filtering, which are the most common preprocessing approaches in solving quality problems. A third preprocessing approach, error correction, is discussed in conjunction with *de novo* transcriptome assembly in Chapter 5.

3.2 SOFTWARE FOR QUALITY CONTROL AND PREPROCESSING

Many tools have been developed for read quality control and preprocessing. Tools for checking read quality include FastQC [2] and PRINSEQ [3], which inspect several quality metrics and provide reports with informative visualizations. The PRINSEQ package also offers filtering and trimming functionality. Other preprocessing tools include Trimmomatic [4], Cutadapt [5], and FastX [6], just to name a few. General introduction to FastQC, PRINSEQ, and Trimmomatic is given here, and their features are discussed in more detail in the context of the different quality problems later in the chapter. The examples use paired-end reads from the GM12892 cell line (sample replicate 2), as described in Chapter 2.

3.2.1 FastQC

FastQC is available as a standalone Java program with a graphical user interface (GUI), and it is also easy to use on command line. It has also been integrated in the Galaxy [7] and Chipster [8] platforms, which provide a GUI to a large number of analysis tools. FastQC is relatively fast, taking only minutes to run for tens of millions of reads. The input files can be FASTQ (uncompressed or gzipped) or SAM/BAM files [9]. In addition to listing the number of reads and their quality encoding, FastQC reports and visualizes information on base quality and content, read length, and k-mer content, and also on the presence of ambiguous bases, over-represented sequences, and duplicates.

The following command produces a quality report and creates a folder reads_fastqc for result files:

```
fastqc reads.fastq.gz
```

The file fastqc_report.html visualizes the information contained in fastqc_data.txt. In addition to reporting several quality metrics, FastQC also gives a judgment on them. The judgment is available as text in summary.txt (pass, warn, fail) and as traffic lights in the html report (Figure 3.1). Note that it is based on general thresholds and might not be

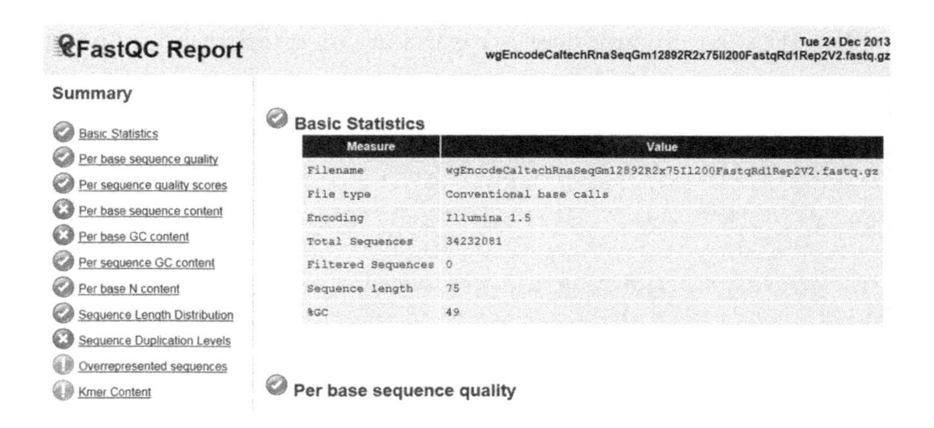

FIGURE 3.1 The beginning of the FastQC quality report offers basic statistics (right) and a judgment on the different quality aspects measured (left).

applicable to your data. For example, the data used here fails the "Sequence Duplication Levels" check, but high duplication levels can be normal for RNA-seq data, as discussed later in this chapter. FastQC also reports some general information about the data, such as the number and length of the reads, and the quality encoding used (note that if you run FastQC using aligned reads, i.e., a BAM file, the number of "reads" reported is actually the number of alignments).

3.2.2 PRINSEQ

PRINSEQ is available as a web application [10], and a standalone version is offered for command line use. It is also available in the Chipster GUI. PRINSEQ quality control reports the number of reads and their length distribution, base quality distribution, sequence complexity, and GC content, as well as the presence of Ns, polyA/T tails, duplicates, and adapters. If problems in any of these are detected, PRINSEQ's trimming and filtering options offer a wide variety of ways of dealing with them. PRINSEQ accepts uncompressed FASTQ, FASTA, and QUAL files. Quality reporting, trimming, and filtering are accomplished with the Perl program prinseq-lite.pl. You can combine many trimming and filtering options in one command. Their processing order does not depend on how you list them in the command, as it is hard-coded in PRINSEQ. The order is described in the help menu, which you can access with

```
prinseq-lite.pl -help
```

PRINSEQ can produce quality reports in either text or html format. In order to create an html report, you need two commands. The first one makes a temporary graph file:

```
prinseq-lite.pl -fastq reads.fastq -phred64 -out_good
null -out_bad null -graph_data graph
```

As we are not performing any preprocessing and hence will not have any accepted or discarded reads, we set the output files for these (`-out_good` and `-out_bad`) to `null`. The qualifier `–phred64` is added because the example data use Illumina's older quality encoding (see below). This first command can take a couple of hours to run. Note that you can request only a subset of quality statistics in order to reduce the memory consumption and running time. For example, adding `-graph_stats ld,gc,qd,ns,pt,ts,de` would skip the sequence complexity and dinucleotide calculation and report only exact duplicates (instead of reporting also 5′ and 3′ duplicates).

The graph file is used for creating the html file. The `-o` parameter gives the file prefix, so this command produces a file QCreport.html.

```
prinseq-graphs.pl -i graph -html_all -o QCreport
```

3.2.3 Trimmomatic

Trimmomatic is a versatile Java-based tool for preprocessing reads. You can use it either on command line or via a GUI in Galaxy or Chipster. Trimmomatic can remove adapters and trim reads in different ways based on quality. It can also filter reads based on quality and length, and convert base qualities from one encoding system into another. Several steps can be performed with one command by listing them in the desired order. Inputs and outputs are FASTQ files, which can be compressed. Trimmomatic is multithreaded, so it runs very fast.

3.3 READ QUALITY ISSUES

3.3.1 Base Quality

Base quality indicates the confidence in the base call. It is expressed in Phred scale, where $\log_{10}$ is taken of the probability that the base is wrong, and multiplied by –10. For example, if there is a 1 in 100 chance that the base is wrong, the quality value is $q = -10*\log_{10}(0.01) = 20$. Quality values typically range from 0 to 40. Instead of numbers, they are encoded with ASCII characters in FASTQ files in order to save space. The current FASTQ files use the so-called Sanger encoding, where the 33rd ASCII character is used as 0. Be aware that Illumina software versions older than 1.8 produced FASTQ files where the 64th ASCII character is used as 0 instead. For details of the different quality encoding systems, see FASTQ format description [1]. If you do not know which quality encoding your data have, FastQC can detect it for you. If you need to convert FASTQ files from one quality encoding into another, Trimmomatic can do that.

Typically base quality values decrease in the later cycles of sequencing. This is easily detected in box plots visualizing the base quality along reads. Both FastQC and PRINSEQ include this kind of plots in their quality reports. Figure 3.2 shows the FastQC per base sequence quality plots for the paired-end reads of our example data (GM12892 cell line replicate 2). As can be seen from the plots, the forward reads have high quality, whereas the reverse reads have lower quality which deteriorates further toward the ends of the reads.

In addition to inspecting the distribution of base qualities per base position, it is informative to check how the reads' mean quality is distributed. This allows one to see if there is a subset of reads which have an overall bad quality. Both FastQC and PRINSEQ can plot the distribution of reads' mean base qualities. Ideally, the majority of reads should have a mean base quality of 25 or higher. As shown in Figure 3.3, both the forward and reverse reads contain about 2 million reads where the base quality is universally bad.

Reads containing low-quality bases can be either filtered or trimmed. Filtering removes the entire read, whereas trimming allows one to remove only the low-quality ends of reads. If you decide to filter or quality trim paired-end reads, choose a tool that is able to preserve the matching order of reads in the output files when a read (or its pair) is removed. This is important when mapping reads to a reference, because aligners expect to find paired reads in the same order in the two files. Note that not only filtering, but also trimming can disturb the order, because some reads get trimmed away totally, and others can become so short that they are discarded because of that.

3.3.1.1 Filtering

Trimmomatic, FastX, and PRINSEQ can filter reads based on quality. The FastX quality filter tool allows one to set a minimum quality value and

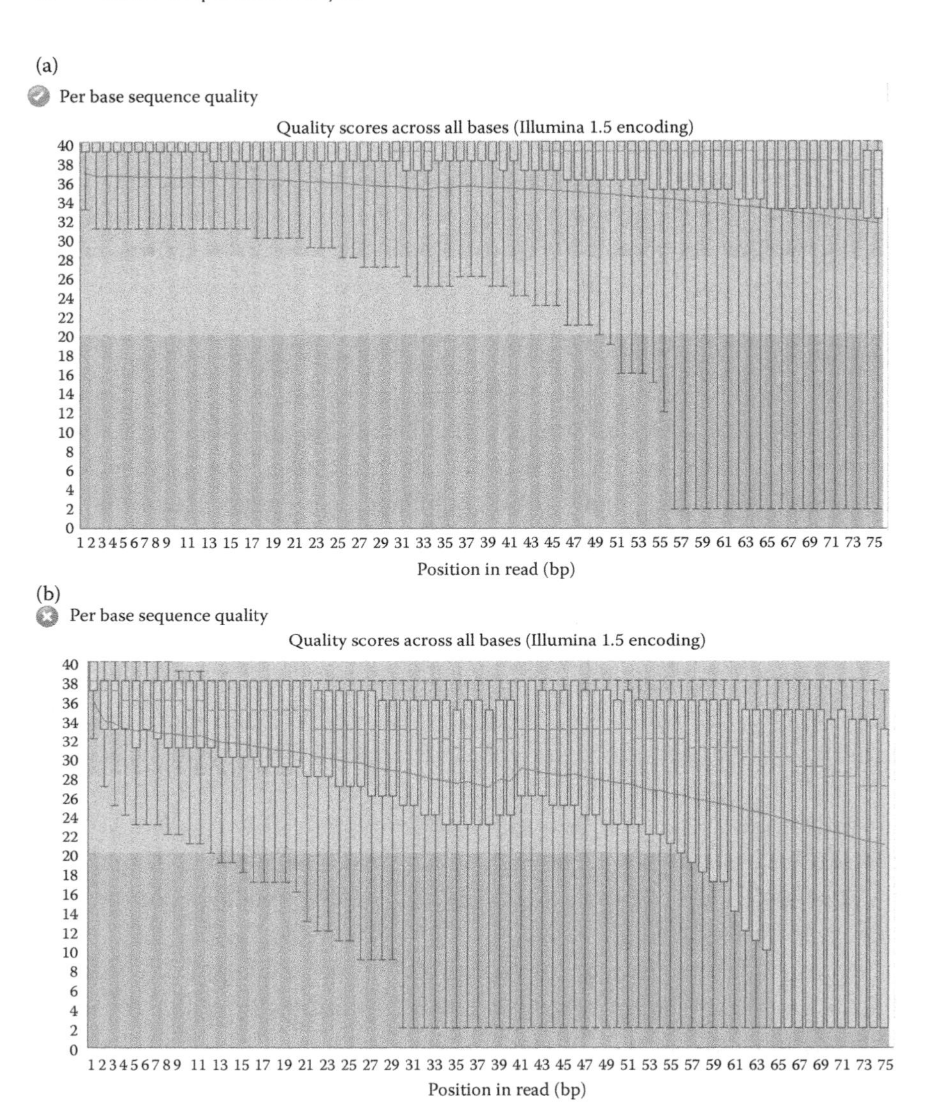

FIGURE 3.2 FastQC per base sequence quality plot for the forward reads (top) and reverse reads (bottom) of the example data. The plot summarizes the base quality at each base position across all reads. The *y*-axis shows the quality scores, and the yellow boxes represent the interquartile range (25–75%) of base quality values for each base position. The red line is the median value and the blue line is the mean. The green, orange, and red background coloring indicate good, reasonable, and poor quality, respectively. FastQC issues a warning if the lower quartile is below 10, or if the median is less than 25 in any of the base positions.

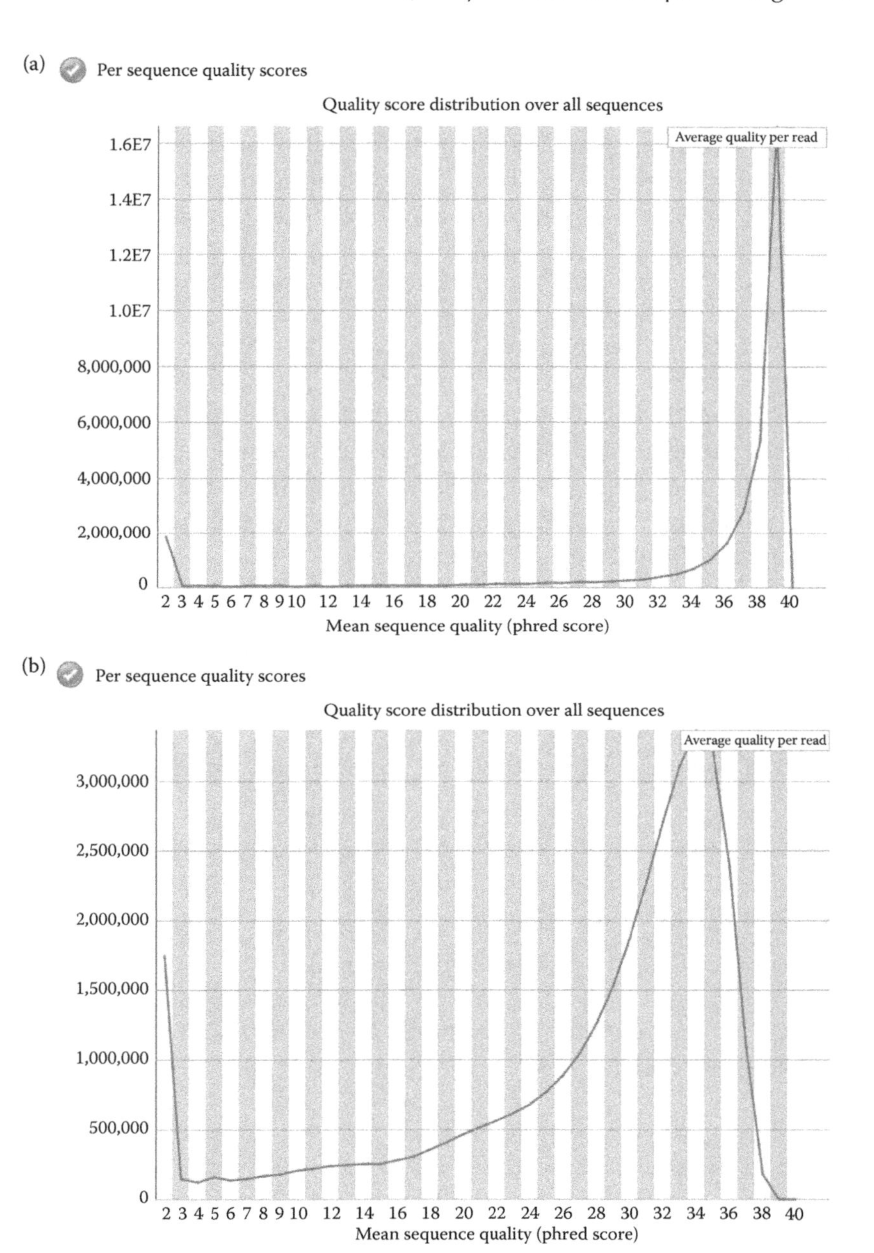

FIGURE 3.3 Per sequence quality score plot by FastQC shows the distribution of reads' mean qualities. Both the forward reads of our example data (top) and the reverse reads (bottom) include a subset of roughly two million reads (~6%), which have a mean quality less than 2. The mean quality of the reverse reads is lower in general.

the percentage of bases that must have this value or higher. PRINSEQ and Trimmomatic can filter based on read's mean base quality, and importantly, they can deal with paired-end reads. The following Trimmomatic command filters out paired-end reads (PE) whose mean base quality is below 20 (AVGQUAL:20). Both the input and the output files can be compressed.

```
java -jar trimmomatic-0.32.jar PE -phred64 reads1.fast
q.gz reads2.fastq.gz paired1.fq.gz unpaired1.fq.gz
paired2.fq.gz unpaired2.fq.gz AVGQUAL:20
```

Trimmomatic checks that the read's pair survived and reports the properly paired reads in files paired1.fq.gz and paired2.fq.gz. The output files unpaired1.fq.gz and unpaired2.fq.gz contain reads, which lost their pair. Here roughly 82% of the pairs survived the filtering as indicated by the screen summary:

```
TrimmomaticPE: Started with arguments: -phred64
reads1.fastq.gz reads2.fastq.gz paired1.fq.gz
unpaired1.fq.gz paired2.fq.gz unpaired2.fq.gz
AVGQUAL:20

Multiple cores found: Using 16 threads

Input Read Pairs: 34232081 Both Surviving:
27981021(81.74%)Forward Only Surviving: 3162984(9.24%)
Reverse Only Surviving: 609823(1.78%)Dropped:
2478253(7.24%)

TrimmomaticPE: Completed successfully
```

The same quality filtering can be performed with PRINSEQ using the following command:

```
prinseq-lite.pl -fastq reads1.fastq -fastq2 reads2.
fastq -phred64 -min_qual_mean 20 -out_good qual_
filtered -out_bad null -no_qual_header -log -verbose
```

The surviving pairs are reported in files qual_filtered_1.fastq and qual_filtered_2.fastq, and the output files qual_filtered_1_singletons.fastq and qual_filtered_2_singletons.fastq contain the reads which lost their pair. The -verbose qualifier allows one to follow the run and view the statistics (available also in the log file). The -no_qual_header tells PRINSEQ to write only

" + " instead of "+ read name" as the header for each read's quality line, in order to reduce the size of the resulting FASTQ files.

3.3.1.2 Trimming

If low-quality bases are detected in the ends of the reads, the simplest way of removing them is to trim reads to a given length or to trim a given number of bases from either end. However, this approach also removes good quality sequence. The loss of sequence can be reduced by considering the quality value of each base: Starting from either the 3′ or 5′ end of the read, a base is trimmed away if its quality is below a user-defined threshold. FastX quality trimmer trims bases from the 3′ end, whereas PRINSEQ and Trimmomatic can trim reads from both ends. As some reads can become very short, trimmers can typically filter out reads which are shorter than a user-defined minimum length. PRINSEQ, Trimmomatic, and Cutadapt have paired-end support, so they can keep the read files synchronized even if a read loses its pair in the trimming process.

The following Trimmomatic command for paired-end reads (`PE`) trims bases from the 3′ end when the base quality is below 20 (`TRAILING:20`) and filters out reads which are shorter than 50 bases after trimming (`MINLEN:50`). The order of the trimming and filtering steps is defined by the command, so it is important to list them in the correct order.

```
java -jar trimmomatic-0.32.jar PE -phred64 reads1.
fastq.gz reads2.fastq.gz paired1.fq.gz unpaired1.fq.gz
paired2.fq.gz unpaired2.fq.gz TRAILING:20 MINLEN:50
```

Here roughly 82% of the pairs survived the trimming and the subsequent length filtering, as indicated by the screen summary:

```
Input Read Pairs: 34232081 Both Surviving: 27992914
(81.77%) Forward Only Surviving: 3114023 (9.10%)
Reverse Only Surviving: 780195 (2.28%)Dropped: 2344949
(6.85%)
```

Instead of looking at the quality one base at a time, trimming can use a sliding window approach, where the base quality in a user-defined window is compared to a given threshold. Trimmomatic slides the window from the beginning (5′ end) of the read, whereas PRINSEQ allows one to decide from which end the scanning should start. Note that sliding the window from the 5′ end *keeps* the beginning of the read until the quality falls below the threshold, while sliding from the 3′ end *cuts* until it reaches a window with good enough quality. As reads can have dips in the quality also in the middle, sliding the window from the 5′ end typically produces shorter reads. Many reads can be lost altogether, if trimming is combined with filtering for user-defined minimum length. The problem is aggravated with paired-end reads, because losing a read leads to removing its pair as well from the paired files. In addition to the scanning direction, PRINSEQ is more flexible also in terms of other settings: while Trimmomatic allows one to set the window size and always uses the mean quality in that window, PRINSEQ also allows one to decide the step size for moving the window, and whether the mean or minimum quality should be compared to the threshold.

The following Trimmomatic command slides a 3-base window from the 5′ end and cuts reads when the mean quality falls below 20 (`SLIDINGWINDOW:3:20`). It also filters out reads which are shorter than 50 bases (`MINLEN:50`) after trimming. As above, surviving pairs are reported in separate files.

```
java -jar trimmomatic-0.32.jar PE -phred64 reads1.
fastq.gz reads2.fastq.gz paired1.fq.gz unpaired1.fq.gz
paired2.fq.gz unpaired2.fq.gz SLIDINGWINDOW:3:20
MINLEN:50
```

Note that only 64.4% of the pairs survive this trimming and filtering as shown by the screen summary below. You can make the trimming gentler by increasing the window size. For example, using a 7-base window would keep 73.4% of the read pairs.

```
Input Read Pairs: 34232081 Both Surviving: 22045360
(64.40%) Forward Only Surviving: 7811189 (22.82%)
Reverse Only Surviving: 607284 (1.77%) Dropped:
3768248 (11.01%)
```

In comparison, the following PRINSEQ command slides a 3-base window from the opposite direction, 3′ end, and trims reads if the mean base quality is less than (`lt`) 20. It also filters out reads which are shorter than 50 bases after trimming. The surviving pairs are reported in files `window_1.fastq` and `window_2.fastq`, and the output files `window_1_singletons.fastq` and `window_2_singletons.fastq` contain the reads which lost their pair.

```
prinseq-lite.pl -phred64 -trim_qual_window 3 -trim_
qual_type mean -trim_qual_right 20 -trim_qual_rule lt
-fastq reads1.fastq -fastq2 reads2.fastq -out_good
window -out_bad null -verbose -min_len 50 -no_qual_
header
```

Now 28,133,789 (81%) pairs survive the trimming and filtering, as indicated by the summary:

```
Input and filter stats:
  Input sequences (file 1): 34,232,081
  Input bases (file 1): 2,567,406,075
  Input mean length(file 1): 75.00
  Input sequences (file 2): 34,232,081
  Input bases (file 2): 2,567,406,075
  Input mean length(file 2): 75.00
  Good sequences (pairs): 28,133,789
  Good bases (pairs): 4,220,068,350
  Good mean length(pairs): 150.00
  Good sequences (singletons file 1): 3,008,972(8.79%)
  Good bases (singletons file 1): 225,672,900
  Good mean length (singletons file 1): 75.00
  Good sequences (singletons file 2): 769,471(2.25%)
  Good bases (singletons file 2): 57,710,325
  Good mean length (singletons file 2): 75.00
  Bad sequences (file 1): 3,089,320(9.02%)
  Bad bases (file 1): 231,699,000
  Bad mean length(file 1): 75.00
  Bad sequences (file 2): 3,008,972(8.79%)
  Bad bases (file 2): 225,672,900
  Bad mean length (file 2): 75.00
  Sequences filtered by specified parameters:
  trim_qual_right: 3330145
  min_len: 50879967
```

An alternative to the sliding window approach is the running sum method, which was originally implemented in the BWA aligner [11] and is hence often called "BWA quality trimming." It scans reads from the right (3′ end), compares the quality of each base to a given threshold, and sums up the differences as it goes. The read is trimmed at the position where the accumulated "badness" is highest. This approach is implemented in the Cutadapt tool.

Finally, Trimmomatic offers an adaptive quality trimming approach called MAXINFO, which aims to balance keeping as long reads as possible with removing erroneous bases. It takes two parameters, target read length and strictness, and trims reads from the 3′ end calculating a score at each base. If a read would become shorter than the target length, a penalty is applied. For longer reads, the penalty from error probability increases and eventually exceeds the bonus of keeping additional bases. One can control this balance by the strictness parameter, which gets a value between 0 and 1 so that higher values favor read correctness. The following MAXINFO trimming command sets target length = 50 and strictness = 0.7.

```
java -jar trimmomatic-0.32.jar PE -phred64 reads1.
fastq.gz reads2.fastq.gz paired1.fq.gz unpaired1.fq.gz
paired2.fq.gz unpaired2.fq.gz MAXINFO:50:0.7 MINLEN:50
```

Roughly 99% of the read pairs survive this trimming and filtering:

```
Input Read Pairs: 34232081 Both Surviving: 33724880
(98.52%) Forward Only Surviving: 63886 (0.19%)Reverse
Only Surviving: 4564(0.01%) Dropped: 438751 (1.28%)
```

If we increase the strictness parameter to 0.8, thereby favoring read correctness over length, the percentage of surviving pairs drops to 82%:

```
Input Read Pairs: 34232081 Both Surviving: 27993319
(81.78%)Forward Only Surviving: 3113077 (9.09%) Reverse
Only Surviving: 780359(2.28%) Dropped: 2345326 (6.85%)
```

3.3.2 Ambiguous Bases

If a base is not identified during sequencing, then it is indicated as N in the read. Assemblers and aligners have different ways of dealing with ambiguous bases: some replace Ns with a random base, while some replace them with a fixed base. As Ns can lead to misassemblies or false mapping, reads with a high number of Ns should be removed. PRINSEQ quality report includes a plot of occurrence of Ns (Figure 3.4), which allows one to see the percentage of Ns per read.

PRINSEQ's filtering functionality allows one to specify the maximum number or percentage of Ns that a read is allowed to have. The following command filters out paired-end reads which have more than two Ns.

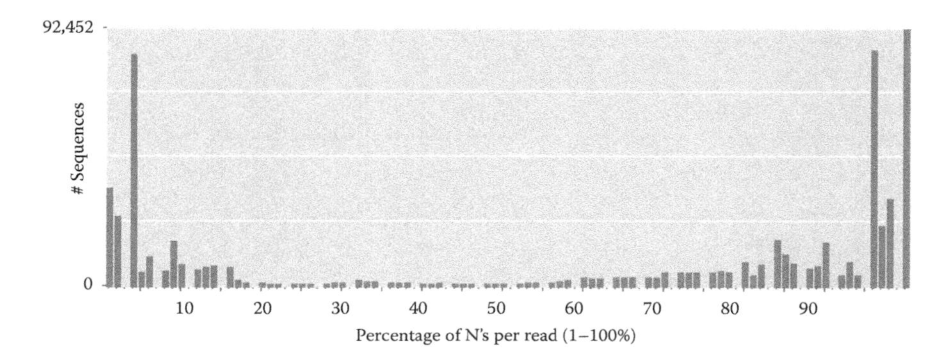

FIGURE 3.4 PRINSEQ report on the occurrence of ambiguous bases (Ns). Two percent of reads contain Ns, and 92,452 reads contain only Ns.

```
prinseq-lite.pl -fastq reads1.fastq -fastq2 reads2.
fastq -ns_max_n 2 -out_good nfiltered -out_bad null
-no_qual_header -log -verbose
```

The surviving pairs are reported in files `nfiltered_1.fastq` and `nfiltered_2.fastq`, and the output files `nfiltered_1_singletons.fastq` and `nfiltered_2 _singletons.fastq` contain the reads which lost their pair. As shown by log, 33,546,906 pairs survived this filtering:

```
Input sequences (file 1): 34,232,081
Input bases (file 1): 2,567,406,075
Input mean length (file 1): 75.00
Input sequences (file 2): 34,232,081
Input bases (file 2): 2,567,406,075
Input mean length(file 2): 75.00
Good sequences (pairs): 33,546,906
Good bases (pairs): 5,032,035,900
Good mean length (pairs): 150.00
Good sequences (singletons file 1): 58,095(0.17%)
Good bases(singletons file 1): 4,357,125
Good mean length (singletons file 1): 75.00
Good sequences (singletons file 2): 141,443(0.41%)
```

```
Good bases (singletons file 2): 10,608,225
Good mean length(singletons file 2): 75.00
Bad sequences (file 1): 627,080(1.83%)
Bad bases (file 1): 47,031,000
Bad mean length(file 1): 75.00
Bad sequences (file 2): 58,095(0.17%)
Bad bases (file 2): 4,357,125
Bad mean length(file 2): 75.00
Sequences filtered by specified parameters:
ns_max_n: 1170812
```

3.3.3 Adapters

As described in Chapter 1, both Illumina and Roche 454 protocols use sequencing adapters, which need to be trimmed away prior to data analysis. Also other tags such as multiplexing identifiers and primers need to be removed. While this sounds trivial, there are some challenges. Firstly, these tags, like any other part of the read, can have sequencing errors, so the trimming software should be able to cope with mismatches, indels, and ambiguous bases. Secondly, when sequencing small RNAs, the reads can extend to the 3′ adapter. The problem with this "read-through" situation is that the adapter in the 3′ end can be partial and therefore difficult to recognize. Finally, if the data come from a public database, the adapter sequence information might not even be available.

If the adapter sequence is not known, TagCleaner [12] can predict it. Also FastQC's k-mer overrepresentation plot and PRINSEQ's Tag Sequence Check plot allow one to detect the presence of adapters. Tools available for adapter removal include Trimmomatic, FastX, TagCleaner, and Cutadapt. Of these, Trimmomatic, TagCleaner, and Cutadapt can cope with mismatches, trim adapters in both ends, and allow the user to specify the minimum overlap of the read and the tag sequence. TagCleaner can also cope with indels and ambiguous bases. Trimmomatic is fast because it scans reads first with short seed sequences and performs the full alignment only for reads that match well with the seeds. It has paired-end support and it can also use the paired-end information for adapter detection. This so-called palindrome approach is based on the fact that "read-through" happens in both directions and the fragment is completely sequenced. Therefore, the reads can be aligned, allowing the detection of partial adapters even down to one base (Figure 3.5). Note that if you combine general, quality, and adapter trimming in one Trimmomatic

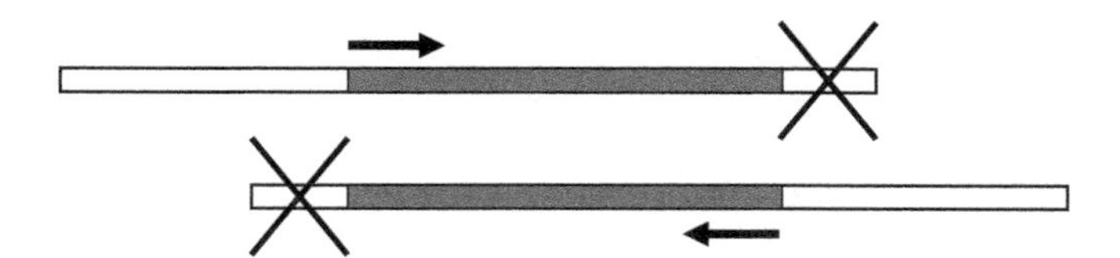

FIGURE 3.5 Palindrome approach of the Trimmomatic software can detect even very short partial adapters in paired-end reads in a "read-through" situation. Two paired-end reads are aligned (forward read on the top and reverse below). Adaptors are white and the insert to be sequenced is black. When the insert is short, sequencing "reads through" it to the 3′ end, resulting in a partial (or full) adapter in that end. Trimmomatic can recognize and remove them (as indicated by the crosses).

command, then you should list the adapter trimming step first, because identifying whole adapters is easier than partial ones.

The example data have already been adapter-trimmed, but the following example command would remove Illumina TruSeq2 adapters from paired-end reads. It allows two mismatches in the seed, palindrome clip threshold is 30, simple clip threshold is 10, the minimum adapter length detected by the palindrome mode is 1, and the reverse read is kept (by default it is removed).

```
java -jar trimmomatic-0.32.jar PE -phred64 reads1.
fastq.gz reads2.fastq.gz paired1.fq.gz unpaired1.fq.gz
paired2.fq.gz unpaired2.fq.gz ILLUMINACLIP:TruSeq2
-PE.fa:2:30:10:1:true
```

3.3.4 Read Length

It is a good practice to check the read length distribution as part of the quality control. This also applies to Illumina reads which are originally of uniform length, because trimming based on quality or adapters can result in very short fragments. Both FastQC and PRINSEQ provide read length distribution plots. Many trimming tools including Trimmomatic, PRINSEQ, Cutadapt, and FastX Quality Trimmer offer the possibility to filter based on read length. The minimum length required depends on the downstream application. Very short reads are difficult to map unambiguously to the genome and longer reads are beneficial also for assembly and quantitation of splice isoforms.

3.3.5 Sequence-Specific Bias and Mismatches Caused by Random Hexamer Priming

During library preparation, RNA is fragmented and random hexamer primers are used to prime the reverse transcription to produce cDNA, the ends of which are eventually sequenced. Therefore, one would expect the reads to start at random locations along the transcripts, and consequently there should not be any base composition bias along the reads. However, it has been shown that priming using random hexamers induces biases in the nucleotide composition at the beginning of RNA-seq reads [13]. This sequence-specific bias affects the expression estimates of genes and isoforms, as the coverage is not uniform along transcripts. Strong biases at different base positions can also be a sign of untrimmed adapters or an overrepresented sequence in the library.

FastQC plots base composition along the reads (Figure 3.6), which should produce a flat line where the amount of each base resembles that of the organism. If the difference between A and T or G and C is bigger than 10% at any read position, FastQC issues a warning.

Sequence-specific bias cannot be removed by filtering or trimming, but the Cufflinks and eXpress packages described in Chapter 6 provide a correction method, which learns the selected sequences from the data and includes this information in the abundance estimation [14].

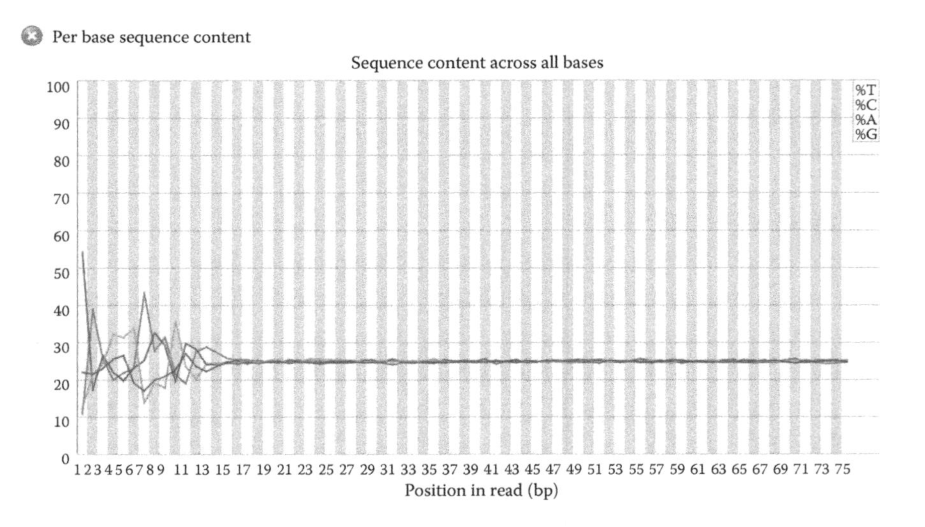

FIGURE 3.6 Per base sequence content plot produced by FastQC. The y-axis shows the percentage of each nucleotide. The first 13 base positions show sequence-specific bias typical for Illumina RNA-seq reads.

Note that in addition to the sequence-specific bias, random hexamer primers have been shown to cause mismatches in the beginning of the Illumina RNA-seq reads [15]. The mismatch rate is highest in the first nucleotide, although up to seven nucleotides can be affected. The mismatch bases have typically good base quality values, so they cannot be detected by this. In order to avoid them, the first base(s) can be trimmed off by Trimmomatic or PRINSEQ.

3.3.6 GC Content

Reads' GC content should follow a normal distribution and center on the GC content of the organism. An unusually shaped distribution or a large shift from the source genome's GC content can indicate a library contamination, which is discussed later in this chapter. Both FastQC and PRINSEQ plot the distribution of reads' mean GC. FastQC also plots GC content per base position, which should produce a flat line at the level of the source genome's GC content. Different GC content at certain base positions indicates a presence of an overrepresented sequence in the library. The previously discussed sequence-specific bias shows also in the GC content plot.

Note that the standard library preparation procedures that employ PCR amplification have been shown to struggle with GC-rich and GC-poor regions. Somewhat unexpectedly, these GC content effects can be sample-specific, which complicates differential expression analysis [16]. The GC bias cannot be removed by preprocessing, but various normalization and correction methods have been proposed to take it into account in the later stages of analysis [14,16].

3.3.7 Duplicates

As discussed earlier, reads are ends of random fragments, so most reads should be unique. In other next-generation sequencing applications, a high level of identical reads can indicate PCR overamplification, but in the context of RNA-seq duplicates are often a natural consequence of sequencing highly expressed transcripts. For differential expression analysis, it is not recommended to remove duplicates, because they would flatten the dynamic range and read counts would not be proportional to the expression level any more. However, if a sparsely covered transcript has a tower of reads in one position, this is likely to indicate a PCR artifact.

Both FastQC and PRINSEQ perform duplicate analysis. FastQC detects only exact duplicates, but PRINSEQ can also detect 5′ and 3′ end duplicates. According to PRINSEQ's quality report, 51.4% of the forward reads

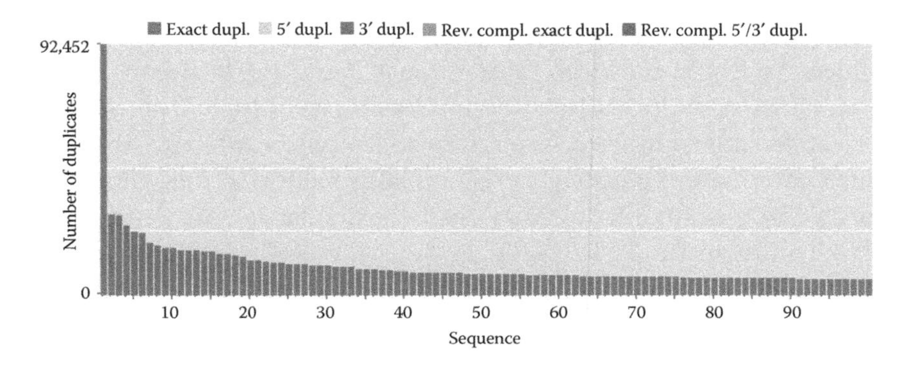

FIGURE 3.7 Extract of PRINSEQ's duplicate report showing the number of duplicates for the 100 most duplicated reads. The most duplicated read has 92,452 copies.

in our example data have exact duplicates. PRINSEQ provides a plot showing how many duplicates the 100 most duplicated reads have, which helps one to determine whether there are many reads with a low level of duplication or a few very highly duplicated reads (Figure 3.7).

PRINSEQ's filtering functionality allows the user to specify how many duplicates a read is allowed to have. FastX Collapser combines identical reads to a single read and keeps count of the reads. These tools work on raw reads and identify duplicates based on sequence. Note that in reality reads from PCR duplicates of the same fragment might not have identical sequence due to sequencing errors, so the sequence-based approach can underestimate the amount of duplicates. Once reads have been aligned to a reference, duplicates can be detected based on identical mapping position rather than sequence content. Toolkits for aligned reads discussed in the next chapter include tools for this. They use the outer genomic mapping coordinates of paired-end reads as an indication of identical fragments. This is not ideal for RNA-seq reads though, because reads from alternative transcripts can have the same outer genomic coordinates but different content due to exon skipping, etc.

If you need to remove duplicates of raw reads, you can use the PRINSEQ package. This example command removes exact duplicate reads (`-derep 1`) which occur more than 100 times (`-derep _ min 101`).

```
prinseq-lite.pl -fastq reads1.fastq -fastq2 reads2.
fastq -derep 1 -derep_min 101 -log -verbose -out_good
dupfiltered -out_bad null -no_qual_header
```

The surviving pairs are reported in files dupfiltered_1.fastq and dupfiltered_2.fastq, and the output files dupfiltered_1_singletons.fastq and dupfiltered_2_singletons.fastq contain the reads which lost their pair. As reported by the log, this command removed 808,295 (2.4%) reads and kept 33,423,786 pairs (96.7%):

```
Input sequences (file 1): 34,232,081
Input bases (file 1): 2,567,406,075
Input mean length (file 1): 75.00
Input sequences (file 2): 34,232,081
Input bases (file 2): 2,567,406,075
Input mean length (file 2): 75.00
Good sequences (pairs): 33,423,786
Good bases (pairs): 5,013,567,900
Good mean length(pairs): 150.00
Good sequences (singletons file 1): 0(0.00%)
Good sequences (singletons file 2): 0(0.00%)
Bad sequences (file 1): 808,295 (2.36%)
Bad bases (file 1): 60,622,125
Bad mean length(file 1): 75.00
Bad sequences (file 2): 0(0.00%)
Sequences filtered by specified parameters:
derep: 808295
```

3.3.8 Sequence Contamination

If you are unlucky, your reads can contain sequences for some contaminating organisms or vectors. As mentioned above, this could be shown in the GC content distribution. PRINSEQ's dinucleotide frequency plot can also give clues about contamination. Probably the most direct way, however, is to align reads to sequences from possible contaminants. The FastQ Screen tool maps reads to user-defined suspects with the Bowtie aligner and summarizes the results in both textual and graphical forms [17]. Alternatively, you could BLAST a random subset of reads to a general nucleotide database.

3.3.9 Low-Complexity Sequences and PolyA Tails

Low-complexity sequences have limited information content and are therefore difficult to map reliably to a reference. For example, they can consist of homopolymer (e.g., AAAAAAAAAA), dinucleotide (e.g., CACACACACA), or trinucleotide (e.g., CATCATCATCAT) repeats. PRINSEQ reports the sequence complexity of reads calculated with two methods: DUST and Entropy. DUST scores range from 0 to 100, homopolymers having the highest value. A read with a DUST score higher than 7 is considered low complexity. Entropy scores are the opposite, so homopolymers have the entropy value of 0, and anything below 50 is considered low complexity. You can filter reads for low complexity with PRINSEQ using the option `-lc-threshold`, and you have to indicate which method that refers to using the option `-lc-method` (`dust/entropy`).

PolyA/T tails are repeats of As or Ts at the end of reads. PRINSEQ reports how many reads contain polyA/T tails of five bases or longer, and what is the distribution of tail lengths. You can trim the tails by giving the minimum tail length to the option `-trim_tail_right` (or `-trim_tail_left`).

QUALITY CONTROL AND PREPROCESSING IN CHIPSTER

- You can produce quality reports with FastQC, PRINSEQ, and FastX using the tools in the Quality control category. For example, select your FASTQ file, the tool "Quality control/Read quality with FastQC," and click "Run."
- PRINSEQ tool for filtering based on quality, Ns, GC content, low complexity, length, and duplicates is available in the Preprocessing category. If you have paired-end reads, give both files at the same time and check in the parameter panel that the forward and reverse reads have been assigned correctly.
- The Preprocessing category offers PRINSEQ, Trimmomatic, FastX, and TagCleaner-based trimming tools for removing low-quality bases, adapters, and polyA tails. It also contains the TagCleaner-based tools "Predict adapters" and "Statistics for adapters." Select your FASTQ file, set the parameters and click "Run." If you have paired-end reads, give both files at the same time, and check in the parameter panel that the forward and reverse reads have been assigned correctly.

3.4 SUMMARY

Taken together, reads can have many quality issues, the importance of which varies. While adapters and sequence contamination need to be removed, base quality issues can be more subtle and duplicates can be just normal. Quality requirements also depend on the subsequent use of reads, so there is not any general quality rule that would fit all situations. For example, sequence-specific bias and GC bias disturb isoform abundance estimation and differential expression analysis, read length is more important for *de novo* assembly and isoform discovery than for differential expression analysis, and aligners differ in their ability to cope with erroneous bases.

There is currently no consensus on what is the optimal base quality threshold for trimming in the context of RNA-seq. Trimming low-quality bases can improve *de novo* assembly and the alignment of reads to a reference, but it also reduces coverage because trimmed reads are shorter and there are less of them. Choosing a quality threshold is therefore a tradeoff between the two. While a comprehensive study on different downstream effects of trimming is still missing, a recent report showed that a very gentle trimming using quality threshold 2 or 5 is optimal for *de novo* transcriptome assembly [18]. A more general study showed that reads trimmed using a quality threshold between 20 and 30 had higher alignment percentage, although the total number of reads was greatly reduced [19].

REFERENCES

1. *FASTQ format description.* Available from: http://en.wikipedia.org/wiki/FASTQ_format.
2. *FastQC.* Available from: http://www.bioinformatics.babraham.ac.uk/projects/fastqc/.
3. Schmieder, R. and Edwards, R. Quality control and preprocessing of metagenomic datasets. *Bioinformatics,* 27(6):863–864, 2011.
4. Bolger, A.M., Lohse, M., and Usadel, B. Trimmomatic: A flexible trimmer for Illumina sequence data. *Bioinformatics,* 2014, doi: 10.1093/bioinformatics/btu170.
5. Martin, M. Cutadapt removes adapter sequences from high-throughput sequencing reads. *EMBnet J,* 17:10–12, 2011.
6. *FASTX-toolkit.* Available from: http://hannonlab.cshl.edu/fastx_toolkit/index.html.
7. Goecks, J., Nekrutenko, A., and Taylor, J. Galaxy: A comprehensive approach for supporting accessible, reproducible, and transparent computational research in the life sciences. *Genome Biol,* 11(8):R86, 2010.
8. Kallio, M.A., Tuimala, J.T., Hupponen, T. et al. Chipster: User-friendly analysis software for microarray and other high-throughput data. *BMC Genomics,* 12:507, 2011.
9. Li, H., Handsaker, B., Wysoker, A. et al. The sequence alignment/Map format and SAM tools. *Bioinformatics,* 25(16):2078–2079, 2009.
10. *PRINSEQ web application.* Available from: http://edwards.sdsu.edu/cgi-bin/prinseq/prinseq.cgi.
11. Li, H. and Durbin, R. Fast and accurate long-read alignment with Burrows–Wheeler transform. *Bioinformatics,* 26(5):589–595, 2010.
12. Schmieder, R., Lim, Y.W., Rohwer, F., and Edwards, R. TagCleaner: Identification and removal of tag sequences from genomic and metagenomic datasets. *BMC Bioinformatics,* 11:341, 2010.

13. Hansen, K.D., Brenner, S.E., and Dudoit, S. Biases in Illumina transcriptome sequencing caused by random hexamer priming. *Nucleic Acids Res*, 38(12):e131, 2010.
14. Roberts, A., Trapnell, C., Donaghey, J., Rinn, J.L., and Pachter, L. Improving RNA-seq expression estimates by correcting for fragment bias. *Genome Biol*, 12(3):R22, 2011.
15. van Gurp, T.P., McIntyre, L.M., and Verhoeven, K.J. Consistent errors in first strand cDNA due to random hexamer mispriming. *PLoS ONE*, 8(12):e85583, 2013.
16. Benjamini, Y. and Speed, T.P. Summarizing and correcting the GC content bias in high-throughput sequencing. *Nucleic Acids Res*, 40(10):e72, 2012.
17. *FastQ Screen*. Available from: http://www.bioinformatics.babraham.ac.uk/projects/fastq_screen/.
18. MacManes, M.D. On the optimal trimming of high-throughput mRNA sequence data. *Front Genet*, 5:13, 2014.
19. Del Fabbro, C., Scalabrin, S., Morgante, M., and Giorgi, F.M. An extensive evaluation of read trimming effects on illumina NGS data analysis. *PLoS ONE*, 8(12):e85024, 2013.

CHAPTER 4

Aligning Reads to Reference

4.1 INTRODUCTION

Alignment means lining up sequences to find out where they are similar and how high the similarity is. Aligning or "mapping" reads to a reference genome or transcriptome allows us to estimate where the read originated from. Mapping reads to genome provides genomic location information, which can be used for discovering new genes and transcripts as described in Chapter 5, and for quantifying expression as described in Chapter 6. If a reference genome is not available, or if you want to quantify only known transcripts, reads can be mapped to a transcriptome instead.

Aligning reads to a reference genome is a challenging task for many reasons: reads are relatively short and there are millions of them, while genomes can be large and contain nonunique sequence such as repeats and pseudogenes lowering the "mappability" of these areas. In addition to this, aligners have to cope with mismatches and indels caused by genomic variation and sequencing errors. Finally, many organisms have introns in their genes, so RNA-seq reads align to genome noncontiguously. Placing spliced reads across introns and determining exon–intron boundaries correctly is difficult, because sequence signals at splice sites are limited and introns can be thousands of bases long.

This chapter covers different types of alignment programs and visualization of aligned reads in genomic context. Tools for alignment statistics and manipulation are introduced as well, while annotation-based quality metrics is discussed in Chapter 6.

4.2 ALIGNMENT PROGRAMS

Tens of alignment programs have been developed, offering various approaches to overcome these challenges. Fonseca et al. [1] provide a comprehensive survey of aligners and update the listing on the web [2]. Typically aligners apply some heuristics and use different indexing schemes to speed up the process. Many tools can consider base quality values when scoring mismatches and they can also make use of the expected distance and relative orientation of paired-end reads. Aligners report the confidence in the mapping location as mapping quality ($Q = -10 \log_{10} P$, where P is the probability that the read originated elsewhere). Mapping quality can depend on several things, but the most important one is uniqueness. Some aligners are able to distribute multimapping reads proportionally to the coverage between the equally matching locations.

Spliced aligners specific for RNA-seq reads use different approaches for aligning spliced reads. This can include performing an initial alignment to discover exon junctions, which then guide the final alignment. If genomic annotation is available, aligners can use it for placing spliced reads. Spliced aligners differ in their alignment yield, splice-detection performance, base-wise accuracy, tolerance for mismatches, and indel detection, as shown by the systematic evaluation performed by Engström et al. [3].

The main consideration when choosing an aligner for RNA-seq studies is whether spliced alignments are needed or not. If the organism does not have introns or if microRNAs were sequenced, it is fine to use contiguous aligners like Bowtie [4] or BWA [5], which were originally developed for DNA. These aligners can also be used if reads are mapped to a transcriptome rather than a genome. However, if RNA-seq reads are mapped to genomes which contain introns, a spliced aligner like TopHat [6], STAR [7], or GSNAP [8] is necessary.

4.2.1 Bowtie

Bowtie is one of the most popular aligners due to its speed and low memory requirement. Here we concentrate on its newer version, Bowtie2, which is particularly good for long reads (from 50 up to thousands of bases) and which can perform gapped alignments for indels. The earlier version, Bowtie1, can be more sensitive for shorter reads, but it does not allow gaps. While Bowtie2 itself is not capable of making spliced alignments, it is used as an alignment engine by the spliced aligner TopHat2. Bowtie2 is also able to stream transcriptome alignments directly to the eXpress quantitation tool [9] as described in Chapter 6.

Bowtie2 achieves its speed and small-memory requirement by indexing the reference genome using an FM index which is based on a Burrows–Wheeler transform method. In order to speed up the alignment process even further, it first narrows down the search space by performing a multiseed alignment. In this initial step, Bowtie2 aligns several small pieces of a read ("seeds") without allowing gaps or ambiguous reference bases. The user can control the seed length and interval and the number of mismatches allowed.

Bowtie2 has two alignment modes, end-to-end and local. The end-to-end mode requires that all bases in the read align, while the local one can trim some bases from one or both ends to maximize the alignment score. The default mode is end-to-end, which is also applied when Bowtie2 is run by TopHat2. In this mode, the best possible alignment score is 0, and penalties are subtracted for each mismatch and gap. A mismatch at a high-quality base receives a higher penalty than a mismatch at a low-confidence base, which might be a sequencing error. The user can select the penalties used and whether base quality information should be taken into account. Rather than setting individual parameter values for the multiseed alignment and the final one, you can use one of the readymade parameter value combinations: very fast, fast, sensitive (default), and very sensitive.

By default, Bowtie2 searches for several alignments until it reaches the limit placed on the search effort and reports the best one. If there are several equally good alignments, it randomly selects one of them, reports the number of alternatives, and lowers the mapping quality value to indicate the lack of confidence in the read's origin. The Bowtie2 example below shows how to align reads to genome. Note that Bowtie2 can be equally well used for aligning reads to transcriptome, and an example of this is given in Chapter 6 in the context of transcript quantitation with eXpress.

Building or Downloading a Reference Index

Bowtie2 reference genome indexes are available for many organisms at the Bowtie2 website [10] and the Illumina iGenomes website [11]. When downloading the index, make sure that you select one for Bowtie2, as indexes for the earlier Bowtie version are different. You can also easily build an index yourself using the `bowtie2-build` command as shown below. Either way, you might like to use genome index/FASTA files from the same provider as GTF files later on, so that chromosome names match (e.g., 1 vs chr1). Ensembl can be a good choice if you are going to quantitate expression with HTSeq later on, as Ensembl GTFs have a correct format for that. GTFs downloaded from iGenomes have additional fields that can be used by the Cuffdiff program.

In our example, we use genome FASTA files from Ensembl to build the Bowtie2 index:

At http://www.ensembl.org/info/data/ftp/index.html, select the organism and the option "DNA." You want the file "dna.toplevel.fa.gz" which contains all the chromosomes in one file (avoid the files "dna_rm" and "dna_sm" which contain repeat masked DNA). Note that this FASTA file contains assembly patches and haplotype sequences in addition to the chromosomes. If you would like to use just the chromosomes, you can download the individual FASTA files for each chromosome and merge them.

Download the file

```
wget ftp://ftp.ensembl.org/pub/release-74/fasta/homo_
sapiens/dna/Homo_sapiens.GRCh37.74.dna.toplevel.fa.gz
```

and unzip it

```
gunzip Homo_sapiens.GRCh37.74.dna.toplevel.fa.gz
```

Build index with `bowtie2-build` command:

```
bowtie2-build -f Homo_sapiens.GRCh37.74.dna.toplevel.
fa GRCh37.74
```

The end part of the command (GRCh37.74) is the index basename that is used in the six bt2 files forming the index:

```
GRCh37.74.1.bt2
GRCh37.74.2.bt2
GRCh37.74.3.bt2
GRCh37.74.4.bt2
GRCh37.74.rev.1.bt2
GRCh37.74.rev.2.bt2
```

The genome FASTA file is not needed by Bowtie2 for the alignment, but you should keep it if the index is going to be used by TopHat2. In that case, you should also rename it to match the index basename:

```
mv Homo_sapiens.GRCh37.74.dna.toplevel.fa GRCh37.74.fa
```

Align Reads to Genome

Bowtie2 accepts FASTQ and FASTA as input files, and they can be compressed. A typical alignment command for single-end reads is shown below. If your sample is split into several input files, they should be separated by commas.

```
bowtie2 -q --phred64 -p 8 --no-unal -x GRCh37.74 -U
reads1.fastq.gz -S reads1aligned.sam
```

Here the reference index basename (–x) is GRCh37.74, and the input file (`-U`) is in FASTQ format (`-q`) which uses the Phred+64 quality encoding (`--phred64`). Note that if your FASTQ files are more recent and have the Phred+33 quality encoding, you should use `--phred33` here. The output is written to a file called reads1aligned.sam (`-S`), and it should not include unaligned reads (`--no-unal`). Eight processors are used simultaneously (`-p 8`) to achieve greater alignment speed. Many other options can be specified too, please see the manual for a thorough description of the different parameters. As indicated by the screen summary, 47.42% of the reads aligned uniquely and the overall alignment rate was 82.68%.

```
34232081 reads; of these:
  34232081(100.00%)were unpaired; of these:
    5928253(17.32%)aligned 0 times
    16232369(47.42%)aligned exactly 1 time
    12071459(35.26%)aligned>1 times
82.68% overall alignment rate
```

The command is similar for paired-end reads, except that now we have two input files (–1 and –2):

```
bowtie2 -q -phred64 -p 8 --no-unal -x GRCh37.74 -1
reads1.fastq.gz -2 reads2.fastq.gz -S paired.sam
```

Note that reads must be in the same order in the two files so that Bowtie2 can handle them as a pair. If you have more files, you should give them in matching order and separated by commas. There are several parameters available specifically for paired-end reads, such as maximum fragment length, whether reads have to map concordantly (with correct relative orientation and distance), and whether unpaired alignments are allowed. As indicated by the screen summary, 39.19% of the read pairs aligned concordantly exactly once, and 26.83% aligned concordantly more than once. The overall alignment rate was 81.67%.

```
34232081 reads; of these:
  34232081(100.00%)were paired; of these:
    11633330(33.98%)aligned concordantly 0 times
    13415597(39.19%)aligned concordantly exactly 1 time
    9183154(26.83%)aligned concordantly >1 times
    ----
    11633330 pairs aligned concordantly 0 times; of these:
      1999775(17.19%)aligned discordantly 1 time
    ----
    9633555 pairs aligned 0 times concordantly or
discordantly; of these:
      19267110 mates make up the pairs; of these:
        12546751(65.12%)aligned 0 times
        4349286(22.57%)aligned exactly 1 time
        2371073(12.31%)aligned >1 times
81.67% overall alignment rate
```

Bowtie2 reports the alignment results in the SAM (Sequence Alignment/Mapped) format, which is the de facto standard for read alignments [12]. In order to save space, SAM can be converted into its binary version BAM as described later in this chapter.

4.2.2 TopHat

The relatively fast and memory efficient TopHat is a commonly used spliced alignment program for RNA-Seq reads. Here we concentrate on TopHat2, which uses Bowtie2 as its alignment engine (Bowtie1 is supported too). It is optimized for reads which are 75 bp or longer. TopHat2 uses a multistep alignment process which starts by aligning reads to transcriptome if genomic annotation is available. This improves alignment accuracy, avoids absorbing reads to pseudogenes, and speeds up the overall alignment process. TopHat2 does not truncate read ends if they do not align. This means low tolerance for mismatches, so reads with low-quality bases might not align well. Finally, TopHat2 can be used to detect genomic translocations, as it can align reads across fusion breakpoints.

The mapping procedure of TopHat2 consists of three major parts, the details of which are listed below: optional transcriptome alignment (step 1), genome alignment (step 2), and spliced alignment (steps 3–6,

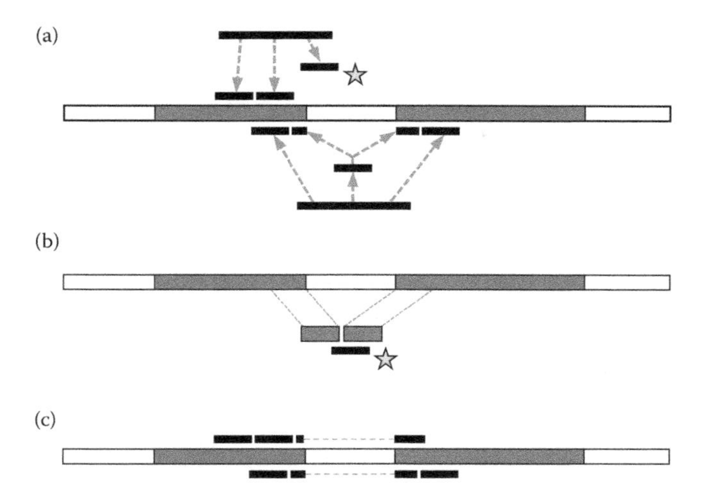

FIGURE 4.1 Spliced alignment procedure of TopHat2. (a) Reads which did not map to the transcriptome or the genome are split into short segments and mapped to the genome again. If TopHat2 finds reads where the left and the right segment map within a user-defined maximum intron size, it maps the whole read to that genomic region in order to find potential splice sites containing known splice signals. (b) Genomic sequences flanking the potential splice sites are concatenated and indexed, and unmapped read segments (marked by a star here) are aligned to this junction flanking index with Bowtie2. (c) Segment alignments are stitched together to form whole read alignments.

shown in Figure 4.1). Paired-end reads are aligned individually first, and then combined to paired-end alignments by taking into account the fragment length and orientation.

1. If annotation information is available, TopHat2 aligns reads to the transcriptome first. It extracts transcript sequences from the Bowtie2 genome index using a GTF/GFF file. Bowtie2 is then used for indexing this virtual transcriptome and aligning reads to it. The transcriptome alignments are converted into (spliced) genomic mappings in the final TopHat2 output.

2. The reads that did not fully align to the transcriptome are aligned to the genome with Bowtie2. At this stage, the reads which map contiguously (to one exon) will be mapped, while multiexon spliced reads will not.

3. The unmapped reads are split into short segments (25 bp by default) and mapped to the genome again (Figure 4.1). If TopHat2 finds reads

where the left and the right segment map within a user-defined maximum intron size, it maps the whole read to that genomic region in order to find potential splice sites containing known splice signals (GT-AG, GC-AG, or AT-AC). TopHat2 also looks for indels and fusion break points at this step.

4. Genomic sequences flanking the potential splice sites are concatenated and indexed, and unmapped read segments are aligned to this junction flanking index with Bowtie2.

5. Segment alignments from steps 3 and 4 are stitched together to form whole read alignments.

6. Reads that extended a few bases into an intron in step 2 are realigned to exons using the new splice site information.

7. In order to decide which alignments to report for multimapping reads, TopHat2 recalculates their alignment score taking into account how many reads support the splice junctions, indels, etc.

Preparing the Reference Indexes

In order to use TopHat2, you need to index the reference genome as described for Bowtie2 earlier in this chapter. TopHat2 also needs the corresponding genomic FASTA file, so do not delete it when the index is ready. If the FASTA file is not available in the same directory as the index files, TopHat2 will create it from the index files in every run, which is a time-consuming process.

If genomic annotations are available in GTF/GFF file format [13], then reads will be aligned to transcriptome first. Ensembl GTFs are available at http://www.ensembl.org/info/data/ftp/index.html by selecting the organism and the option "GTF." You can prepare the transcriptome index beforehand to save time in each subsequent alignment run:

```
tophat2 -G GRCh37.74.gtf --transcriptome-index=
GRCh37.74.tr GRCh37.74
```

Here we use the annotation file GRCh37.74.gtf and the Bowtie2 genome index GRCh37.74 to build a Bowtie2 transcriptome index which has the basename GRCh37.74.tr. Note that the chromosome names in the GTF file and in the genome index must match. Bowtie2 has to be on the path, because TopHat2 uses it to build the index. The following files are created:

```
GRCh37.74.tr.1.bt2
GRCh37.74.tr.2.bt2
GRCh37.74.tr.3.bt2
GRCh37.74.tr.4.bt2
GRCh37.74.tr.fa
GRCh37.74.tr.fa.tlst
GRCh37.74.tr.gff
GRCh37.74.tr.rev.1.bt2
GRCh37.74.tr.rev.2.bt2
GRCh37.74.tr.ver
```

Aligning the Reads

TopHat2 accepts both FASTQ and FASTA files as input. Read files can be compressed (.gz), but tarballs (.tgz or .tar.gz) need to be opened to separate files. The examples below show separate alignment commands for single-end and paired-end reads, but TopHat2 can also combine single-end reads in a paired-end alignment if needed.

The following two alternative commands align single-end reads. In both cases, reads are aligned to the human reference genome (index basename GRCh37.74), but the first command uses a premade transcriptome index, while the second one builds the transcriptome index on the fly using the GTF file. If you have several read files, separate them with commas. Note that both Bowtie2 and SAMtools have to be on the path, because TopHat2 uses these packages internally.

```
tophat2 -o outputFolder --transcriptome-
index=GRCh37.74.tr -p 8 --phred64-quals GRCh37.74
reads1.fastq.gz
```

or

```
tophat2 -o outputFolder -G GRCh37.74.gtf -p 8
--phred64-quals GRCh37.74 reads1.fastq.gz
```

TopHat2 assumes that the base quality encoding is Sanger (phred+33), so we have to add the qualifier `--phred64-quals` to indicate that the example data come from an earlier Illumina version (`--solexa1.3-qual` would work as well). Here eight processors are used simultaneously

(`-p 8`) to speed up the process. Note that if your data were produced with a strand-specific protocol, you have to set the `--library-type` parameter accordingly (the default is unstranded). TopHat2 has many more alignment and reporting options, for example, you can align reads to the transcriptome only (`-T`), or change the maximum number of alignments reported per read (`-g`) which is 20 by default.

The align_summary.txt indicates that 79.3% of the reads mapped:

```
Reads:
      Input     :  34232081
       Mapped   :  27140089(79.3% of input)
        of these:   1612317 (5.9%)have multiple
alignments (2771 have >20)
79.3% overall read mapping rate.
```

The alignment command for paired-end reads is shown below. Note that the order of the reads in the two files has to match so that TopHat2 can pair them correctly. If you have several read files, separate them with commas and enter them in the same order, leaving an empty space between the two sets.

```
tophat2 -o outputFolder --transcriptome-
index=GRCh37.74.tr -p 8 --phred64-quals GRCh37.74
reads1.fastq.gz reads2.fastq.gz
```

TopHat parameters specific to paired-end alignments include the expected inner distance between the paired reads (`-r`), which you should set according to your data. The default value 50 is suitable here, because the insert size of the example data is 200 and reads are 75 bases long (200 - 2 * 75 = 50). You can also request that a pair has to map concordantly, that is, with the expected orientation and distance (`--no-discordant`). If TopHat cannot map a pair together, it will map the reads separately, but you can disable this default behavior (`--no-mixed`).

Now the following summary is produced:

```
Left reads:
          Input     :  34232081
           Mapped   :  27143093(79.3% of input)
            of these:   1014796 (3.7%)have multiple
alignments (3621 have >20)
```

```
Right reads:
     Input       :  34232081
      Mapped     :  22600062 (66.0% of input)
        of these:      759539  (3.4%)have multiple
alignments (3193 have >20)
72.7% overall read mapping rate.

Aligned pairs:  21229613
     of these:      702920 (3.3%)have multiple alignments
                    336032 (1.6%)are discordant alignments
61.0% concordant pair alignment rate.
```

TopHat produces several result files:

- accepted_hits.bam contains the alignments in BAM format. The alignments are sorted according to chromosomal coordinates.
- junctions.bed contains the discovered exon junctions in BED [14] format. A junction consists of two blocks, where each block is as long as the longest overhang of any read spanning the junction. The score is the number of alignments spanning the junction.
- insertions.bed contains the discovered insertions. chromLeft refers to the last genomic base before the insertion.
- deletions.bed contains the discovered deletions. chromLeft refers to the first genomic base of the deletion.
- align_summary.txt reports the alignment rate and how many reads and pairs had multiple alignments.

4.2.3 STAR

STAR (Spliced Transcripts Alignment to a Reference) is a relatively new spliced alignment program which runs very fast. The tradeoff is that it needs considerably more memory than TopHat, for example. The STAR manual (as of February 11, 2013) states that 31 GB RAM is "enough for human and mouse," but it is also possible to run it with 16 GB for the human genome if the reference index is built in the proper way (see below). While STAR is particularly known for its speed, it also has many other advantages. It can perform an unbiased search for splice junctions because it does not need any prior information on their locations, sequence signals, or intron length. STAR can align a read containing any number of splice junctions, indels, and mismatches, and it can cope with poor-quality ends. Finally, it can map long reads and even full-length mRNA, which is required as read lengths are increasing.

The benefits of STAR are largely based on the so-called "maximum mappable length" approach. STAR splits a read into pieces (which are by default 50 bases long) and finds the best portion that can be mapped for each piece. It then maps the remaining portion, which can be far away in the case of a splice junction. This sequential maximum mappable seed search looks for exact matches and uses the genome in the form of uncompressed suffix arrays. The second step of STAR stitches the seeds together within a given genomic window and allows for mismatches, indels, and splice junctions. The seeds from read pairs are handled concurrently at this step in order to increase sensitivity.

STAR can find spliced junctions *de novo*, but you can also supply it with junction annotations when building the reference index. In this case, a user-defined number of exonic bases from both the splice donor and acceptor sites are combined, and these sequences are added to the genome sequence. During mapping, reads are aligned to both the genome sequence and the splice site sequences. If a read maps to the splice sequence and crosses the junction in it, the coordinates of this mapping are combined with the genomic ones.

Building or Downloading a Reference Index

You need to build or otherwise obtain a reference index for your genome of interest before running STAR. For some genomes (human, mouse, sheep, and chicken), there are prebuilt STAR reference indexes available for download (ftp://ftp2.cshl.edu/gingeraslab/tracks/STARrelease/STARgenomes/). There are a few different indexes for the human genome built with different use cases in mind. In particular, the one where the name contains the word "sparse" is built for use with less memory. If you want to build your own index, you need to give the following type of STAR command:

```
STAR --runMode genomeGenerate --genomeDir /path/to/
GenomeDir --genomeFastaFiles fasta1 fasta2
--sjdbFileChrStartEnd annotation.gtf.sjdb
--sjdbOverhang 74 --runThreadN 8
```

The `--genomeDir` option indicates the directory where the reference index (consisting of binary genome sequence, suffix array files, and some auxiliary files) will be located. The --genomeFastaFiles lists the reference sequence FASTA files to be indexed. The indexing process can be run in a multithreaded fashion using the --runThreadN option. If you want to use a splice junction annotation in the mapping (which is usually a good idea), you need to provide a splice junction reference file when you construct the reference index. The example command uses the parameter --sjdbFileChrStartEnd to supply the file annotation.gtf.sjdb, which contains the genomic coordinates of introns in a format defined in the STAR manual. Such a file for the human reference genome hg19 can be downloaded from the link specified above. You can alternatively use a GTF file with a parameter --sjdbGTFfile. In both cases, you have to use -sjdbOverhang parameter to define how long sequences from the known donor and acceptor sites should be used when constructing the reference index. Ideally this value should be set to read length –1, so the example above is assuming 75 bp reads. If you have reads of varying length, using a large value is safer. If you need to decrease the amount of memory needed to run STAR, you can try to build the reference index with a higher value for the --genomeSAsparseD option (the default value is 1). This will use a sparser suffix array, which lowers the memory requirements at the expense of alignment speed.

Mapping

The following mapping command for STAR uses a prebuilt, splice-junction-annotated human genome index, which has been downloaded from the STAR home page (see the link above):

```
STAR --genomeDir hg19_Gencode14.overhang75
--readFilesIn reads1.fastq.gz reads2.fastq.gz
--readFilesCommand zcat --outSAMstrandField
intronMotif --runThreadN 8
```

The --genomeDir option should point to the reference index directory that you have built or downloaded according to the instructions above. Next, after --readFilesIn, specify the FASTQ file(s). These can be compressed, but in that case you need to specify a command to unpack the particular compression format as an argument to the --readFilesCommand option (here zcat is used). If you have several read files, separate them by commas and leave an empty space before listing the mate files in a matching order. The parameter --outSAMstrandFieldintronMotif adds the SAM strand attribute XS that is needed downstream by the Cufflinks program, in case you are planning to use it. There are many other parameters controlling various aspects of STAR's behavior as described in the manual. For example, you might like to filter out alignments which contain more than a given number of mismatches, or which contain splice junctions supported by too few reads.

Output

As of December 2013, STAR outputs at least the following files:

- Aligned.out.sam—Alignments in SAM format (reads that were not aligned are not included).
- SJ.out.tab—A tab-delimited file containing information on alignments to splice junctions.
- Log.out, Log.final.out, Log.progress.out—As the names indicate, these are log files providing various information about how the run is proceeding. It is often of interest to look at the Log.final.out file shown below, because it provides useful mapping statistics. Note that the number of reads and read length combines the read pairs.

```
                         Started job on | Feb 12 11:32:58
                     Started mapping on | Feb 12 11:46:52
                            Finished on | Feb 12 11:51:09
Mapping speed, Million of reads per hour | 479.52
                  Number of input reads | 34232081
              Average input read length | 150
                            UNIQUE READS:
           Uniquely mapped reads number | 27113906
                 Uniquely mapped reads% | 79.21%
                  Average mapped length | 147.51
               Number of splices: Total | 12176905
     Number of splices: Annotated (sjdb) | 12049801
               Number of splices: GT/AG | 12070507
               Number of splices: GC/AG | 78264
               Number of splices: AT/AC | 9359
        Number of splices: Non-canonical | 18775
               Mismatch rate per base,% | 1.04%
                  Deletion rate per base | 0.01%
                 Deletion average length | 2.20
                 Insertion rate per base | 0.02%
                Insertion average length | 1.85
```

```
                    MULTI-MAPPING READS:
 Number of reads mapped to multiple loci | 1376440
      % of reads mapped to multiple loci | 4.02%
 Number of reads mapped to too many loci | 7662
      % of reads mapped to too many loci | 0.02%
                         UNMAPPED READS:
% of reads unmapped: too many mismatches | 0.00%
          % of reads unmapped: too short | 15.70%
              % of reads unmapped: other | 1.05%
```

ALIGNING READS TO REFERENCE IN CHIPSTER

Chipster offers Bowtie2, BWA, and TopHat for aligning reads to a reference, and separate tools are available for single and paired-end reads.

- Select your read files (FASTQ) and one of the tools in the Alignment category. In the parameter panel, select the right reference and alignment options and check that the files have been assigned correctly in the case of paired-end reads and/or own GTF or reference FASTA file.
- The result files are always coordinate-sorted and indexed BAM files.

4.3 ALIGNMENT STATISTICS AND UTILITIES FOR MANIPULATING ALIGNMENT FILES

The SAM/BAM files produced by aligners typically need some processing, such as SAM/BAM conversion, sorting, indexing, or merging. Two major packages are available for these tasks: SAMtools [12] and its Java implementation Picard [15]. Picard has more tools and is also stricter than SAMtools when validating files. Here we focus on some commonly used SAMtools commands.

- Convert SAM into BAM. Storing alignments in the BAM format saves space and many downstream tools use BAM format rather than SAM. Here we specify that input is SAM (-S), output is BAM (-b), and the output file should be named alignments.bam (-o).

```
samtools view -bS -o alignments.bam input.sam
```

- Convert BAM into SAM and include the header information (-h). The header lines start with the "@" sign and contain information about reference sequence names and lengths (@SQ), what program created the file (@PG), and whether and how the file is sorted (@HD).

```
samtools view -h -o alignments.sam input.bam
```

- Retrieve just the header (-H).

```
samtools view -H alignments.bam
```

- Sort alignments in BAM by chromosomal coordinates or by read names (-n). Coordinate sorting is required by genome browsers and some analysis tools, while name sorting is often required by expression quantitation tools.

```
samtools sort alignments.bam alignments.sorted

samtools sort -n alignments.bam alignments.
namesorted
```

- Note that SAMtools can work on a stream, so it is possible to combine commands with Unix pipes in order to avoid large intermediate files. For example, the following command converts Bowtie2's SAM output into BAM and sorts it by chromosomal coordinates, producing a file alignments.sorted.bam:

```
bowtie2 -q --phred64 -p 4 -x GRCh37.74 -U reads1.
fq | samtools view -bS - | samtools sort -
alignments.sorted
```

- Index coordinate-sorted BAM files. Indexing enables fast retrieval of alignments, and it is required by genome browsers and some downstream tools. The following command produces an index file alignments.sorted.bam.bai:

```
samtools index alignments.sorted.bam
```

- Make a subset of alignments by specifying a certain chromosome or chromosomal region (here we extract alignments to chromosome 18). This command requires that an index file is present.

```
samtools view -b -o alignments.18.bam alignments.
bam 18
```

- List how many reads map to each chromosome. This command requires that an index file is present.

  ```
  samtools idxstats alignments.sorted.bam
  ```

- Filter alignments based on mapping quality. The following keeps alignments which have mapping quality higher than 30:

  ```
  samtools view -b -q 30 -o alignments_MQmin30.bam
  alignments.bam
  ```

- Filter alignments based on values in the SAM flag field. The –F option filters out reads which have the given flag value (here 4 which means unmapped reads), and the –f option keeps reads with the given flag value (here 2 which means that a read is mapped in a proper pair). For details of the flag values, please see the SAM specification [12].

  ```
  samtools view -b -F 4 -o alignments.mapped_only.
  bam alignments.bam
  ```

  ```
  samtools view -b -f 2 -o properly_paired_reads.bam
  alignments.bam
  ```

- Obtain mapping statistics based on the flag field.

  ```
  samtools flagstat alignment.bam
  ```

The report contains basic information such as the number of mapped reads and properly paired reads, and how many mates map to a different chromosome:

```
52841623 + 0 in total (QC-passed reads + QC-failed reads)
0 + 0 duplicates
52841623 + 0 mapped (100.00%:-nan%)
52841623 + 0 paired in sequencing
28919461 + 0 read1
23922162 + 0 read2
42664064 + 0 properly paired (80.74%:-nan%)
44904884 + 0 with itself and mate mapped
7936739 + 0 singletons (15.02%:-nan%)
999152 + 0 with mate mapped to a different chr
357082 + 0 with mate mapped to a different chr (mapQ >=5)
```

Alignment statistics can also be obtained with the RseQC package [16], which is covered in more detail in Chapter 6 in the context of annotation-based quality metrics. RseQC consists of several Pythons scripts which check alignment metrics such as how many reads aligned, what proportion of them aligned uniquely, what is the inner distance distribution, and what proportion of pairs map to exactly the same location. The latter can indicate that the reads stem from identical fragments, possibly due to PCR over-amplification. Basic alignment statistics can be obtained with the bam_stat.py tool:

```
python bam_stat.py -i accepted_hits.bam
```

It produces the following table, where reads are considered unique if their mapping quality is more than 30 (you can change the threshold by adding parameter –q).

```
#==================================================
#All numbers are READ count
#==================================================
Totalrecords:                               52841623

QCfailed:                                   0
Optical/PCR duplicate:                      0
Non primary hits                            3098468
Unmapped reads:                             0
mapq<mapq_cut(non-unique):                  1774335

mapq>=mapq_cut(unique):                     47968820
Read-1:                                     26128297
Read-2:                                     21840523
Reads map to'+':                            24085239
Reads map to'-':                            23883581
Non-splice reads:                           35970095
Splice reads:                               11998725
Reads mapped in proper pairs:               39702036
Proper-paired reads map to different chrom:0
```

SAM/BAM MANIPULATION AND ALIGNMENT STATISTICS IN CHIPSTER

- Chipster has many SAMtools-based tools in the Utilities category. They convert SAM into BAM and vice versa, sort, index, subset, and merge BAM files, count alignments per chromosomes and in total, and create a consensus sequence out of alignments. Some tools require a BAM index file as indicated. Select both the BAM file and the index file and check in the parameter panel that the files have been assigned correctly.

- RseQC is available in the Quality control category. It reports BAM statistics, inner distance distribution, and information on strandedness, in addition to the annotation-based quality metrics discussed in Chapter 6.

4.4 VISUALIZING READS IN GENOMIC CONTEXT

Visualizing aligned reads in a genomic context can serve many purposes and is highly recommended. You can visualize the structure of novel transcripts, judge the support for novel junctions, check the coverage of different exons and whether there are "towers" of duplicate reads, spot indels and SNPs, etc. Importantly, you can compare your data with reference annotations.

Several genome browsers are able to visualize high-throughput sequencing data, including the Integrative Genomics Viewer IGV [17], JBrowse [18], Tablet [19], and the UCSC [20] and Chipster genome browsers. These browsers offer a lot of functionalities, and describing them all is beyond the scope of this book. Instead, we recommend reading the special issue of Briefings in Bioinformatics on next-generation sequencing visualization [21], which provides informative articles on several genome browsers. Chapter 2 contains screen shots of the IGV and Chipster genome browsers with RNA-seq data.

As the Chipster software is used in the examples throughout the book, we give a brief introduction to its genome browser here. Chipster visualizes data in the context of Ensembl annotations and supports several file formats including BAM, BED, GTF, VCF, and tsv. Users can zoom in to the nucleotide level, highlight differences from the reference sequence, and view automatically calculated coverage (either total or strand-specific). For BED files, it is possible to visualize also the score. Importantly, different kinds of data can be visualized together. For example, you can view RNA-seq data and copy number of aberrations measured by microarrays side by side. As Chipster genome browser is integrated with a comprehensive analysis environment, you do not need to export and import data to an external application. Of course you can import BAM files to Chipster if you want to use it only for visualization purposes. In that case, your files are automatically sorted and indexed during the import.

VISUALIZING READS IN GENOMIC CONTEXT WITH CHIPSTER

As an example, let us visualize the TopHat2 result files accepted_hits.bam and deletions.bed.

- You can use the BED file as a navigation aid, so detach it first to a separate window: Double click on the file to open it in a spreadsheet view and click "Detach."
- Select the BAM and BED files and the visualization method "Genome browser" in the Visualization panel and maximize the panel for larger viewing area.
- Select hg19 from the Genome pull-down menu and click "Go." You can zoom in and out using the mouse wheel and change the coverage scale if needed.
- Use the detached BED file to inspect the list of deletions efficiently: Click on the start coordinate of a deletion (column 1) and the browser will move to that location. You can also sort the BED file by the score (number of reads supporting the deletion) by clicking on the column 4 title.

4.5 SUMMARY

Mapping millions of RNA-seq reads to a reference genome is a computationally demanding task, and aligners typically use different reference indexing schemes to speed up the process. Many organisms contain introns, so a spliced aligner is required in order to map reads to genome noncontiguously. Aligners also have to support mismatches and indels in order to cope with genomic variants and sequencing errors and take base quality into account when scoring them. Instead of mapping reads to a genome, you can also map them to a transcriptome. This is the only way for organisms, which do not have a reference genome available.

The choice of the aligner depends on the organism and the goal of the experiment. For example, if spliced alignments are not required and accuracy is important, BWA might be a good choice. If speed is more important, Bowtie2 is recommended. If the organism has introns and a nearly complete reference annotation, TopHat2 can produce good spliced alignments.

On the other hand, STAR copes better with mismatches, runs faster, produces more alignments, and can detect splice junctions in an unbiased manner.

Alignment files can be manipulated with various utilities such as SAMtools and Picard, which allow, for example, efficient retrieval of reads which map to a certain region or which map uniquely. Tools like RseQC provide important quality information on aligned reads. Several genome browsers are available for visualizing alignments in genomic context. This is highly recommended, because nothing beats the human eye in detecting interesting patterns in the data.

REFERENCES

1. Fonseca N.A., Rung J., Brazma A., and Marioni J.C. Tools for mapping high-throughput sequencing data. *Bioinformatics* **28**(24):3169–3177, 2012.
2. *Updated listing of mappers*. Available from: http://wwwdev.ebi.ac.uk/fg/hts_mappers/.
3. Engström P.G., Steijger T., Sipos B. et al. Systematic evaluation of spliced alignment programs for RNA-seq data. *Nat Methods* **10**(12):1185–1191, 2013.
4. Langmead B. and Salzberg S.L. Fast gapped-read alignment with Bowtie2. *Nat Methods* **9**(4):357–359, 2012.
5. Li H. and Durbin R. Fast and accurate long-read alignment with Burrows–Wheeler transform. *Bioinformatics* **26**(5):589–595, 2010.
6. Kim D., Pertea G., Trapnell C. et al. TopHat2: Accurate alignment of transcriptomes in the presence of insertions, deletions and gene fusions. *Genome Biol* **14**(4):R36, 2013.
7. Dobin A., Davis C.A., Schlesinger F., et al. STAR: Ultrafast universal RNA-seq aligner. *Bioinformatics* **29**(1):15–21, 2013.
8. Wu T.D. and Nacu S. Fast and SNP-tolerant detection of complex variants and splicing in short reads. *Bioinformatics* **26**(7):873–881, 2010.
9. Roberts A. and Pachter L. Streaming fragment assignment for real-time analysis of sequencing experiments. *Nat Methods* **10**(1):71–73, 2013.
10. *Bowtie2*. Available from: http://bowtie-bio.sourceforge.net/bowtie2/index.shtml.
11. *iGenomes*. Available from: http://support.illumina.com/sequencing/sequencing_software/igenome.ilmn.
12. Li H., Handsaker B., Wysoker A. et al. The sequence alignment/map format and SAMtools. *Bioinformatics* **25**(16):2078–2079, 2009.
13. *GFF/GTF file format description*. Available from: http://genome.ucsc.edu/FAQ/FAQformat.html#format3.
14. *BED file format description*. Available from: http://genome.ucsc.edu/FAQ/FAQformat.html#format1.
15. *Picard*. Available from: http://picard.sourceforge.net/.

16. Wang L., Wang S., and Li W. RSeQC: Quality control of RNA-seq experiments. *Bioinformatics* **28**(16):2184–2185, 2012.
17. Thorvaldsdottir H., Robinson J.T., and Mesirov J.P. Integrative Genomics Viewer (IGV): High-performance genomics data visualization and exploration. *Brief Bioinform* **14**(2):178–192, 2013.
18. Westesson O., Skinner M., and Holmes I. Visualizing next-generation sequencing data with JBrowse. *Brief Bioinform* **14**(2):172–177, 2013.
19. Milne I., Stephen G., Bayer M. et al. Using Tablet for visual exploration of second-generation sequencing data. *Brief Bioinform* **14**(2):193–202, 2013.
20. Kuhn R.M., Haussler D., and Kent W.J. The UCSC genome browser and associated tools. *Brief Bioinform* **14**(2):144–161, 2013.
21. Special Issue: Next generation sequencing visualization. *Brief Bioinform* **14**(12), 2013.

CHAPTER 5

Transcriptome Assembly

5.1 INTRODUCTION

The goal of RNA-seq assembly is to reconstruct full-length transcripts based on sequence reads. Owing to limitations in second-generation sequencing technology, only relatively short fragments can be sequenced as a single unit. Although there are promising methods resulting from the third-generation sequencing technology, such as PacBio of Pacific Biosciences, which allows for sequencing of single molecules of several kilobases in length, they are not at the moment routinely used in transcriptome sequencing. Therefore, in practice, in order to get full-length transcript sequences, one must build them from small overlapping fragments. In principle, there are two ways of doing this. If there is a reference genome available, it can be utilized to guide the assembly. RNA-seq reads are first mapped on the genome and the assembly task consists of solving which mapped reads correspond to which transcripts. The alternative approach is to perform *de novo* assembly which does not utilize any external information. In the absence of a reference genome, the assembly is based on utilizing sequence similarity between the RNA-seq reads. Both these approaches can be formulated as a computational problem which includes finding a set of paths in a graph. Owing to the combinatorial nature of the problem, there is an astronomical number of possible solutions, even in a relatively small assembly task. Enumerating all possible solutions to find the global optimum is simply not possible and therefore various heuristics and approximations are used during the assembly process.

Transcriptome assembly is different from genome assembly. In genome assembly, the read coverage is usually more uniform (excluding biases depending on the library preparation and sequencing technology). Deviation from uniform sequence depth in genome assembly indicates the presence of repeats. In contrast, with RNA-seq data, the abundance of gene expression can vary several magnitudes between genes and also different isoforms of the same gene can be expressed at different levels. Although this can actually be utilized in transcript assembly in detecting and constructing different isoforms, highly different abundances between the genes also introduce challenges. It requires more sequencing depth to represent less abundant genes and rare events. In order to balance abundance differences between the genes, there are wet laboratory procedures for library normalization. Description of such methods is beyond the scope of this book, but it is good to keep in mind that the quality of assembly consists of the combination of data and computational methods. Since sequencing technology only converts the content of an RNA-seq library into a digital form, library preparation is a key element in obtaining good quality data. Garbage in–garbage out applies to both sequencing and assembly. Quality control of data should be done before any assembly.

For this chapter, we have selected two software packages for mapping-based assembly and two software packages for *de novo* assembly. All of them are noncommercial and publicly available. Like using any computational methods, it is good to be aware that the output of the assembly depends on the combination of the data and the method. Typically, each method involves parameters which can be tuned and therefore, depending on the method and the parameters, the output of the assembly can vary considerably even when using the same data.

The chapter starts with the description of the assembly problem and the methodology used for solving it. Each of the four selected software packages is then introduced and their usage is demonstrated with the same data set. The data set is from the ENCODE project, and it includes paired-end reads of one individual. In order to limit the data size, only reads which have been mapped on human chromosome 18 are used in examples. Paired-end sequence reads were extracted from the file "wgEncodeCaltechRnaSeqH1hescR2x75Il200AlignsRep1V2.bam" (http://hgdownload.cse.ucsc.edu/goldenPath/hg19/encodeDCC/wgEncodeCaltechRnaSeq/). The resulting data set is small, containing only 344,000 read pairs, so running the assemblers should not take long time.

5.2 METHODS

The roots of RNA-seq assembly can be traced to the early days of expressed sequence tag (EST) sequencing in the beginning of the 1990s [1]. Processing of ESTs involved clustering and assembly [2]. Clustering meant grouping similar EST reads together by calculating pairwise overlaps. Cluster membership could be defined, for example, if there was 95% identity in an overlap longer than 40 bp with another sequence. After clustering of reads, assembly was carried out separately within each cluster. Although the details have changed, these two steps still constitute the main steps of the transcript assembly process: (1) finding the reads which belong to the same locus and (2) constructing the graph representing the transcripts within each locus. One big difference between the ESTs and present-day RNA-seq data is that typically ESTs represented only fragments and partial transcripts, but the nature of today's high-throughput data enables the representation of full-length transcripts. Although single reads of currently used second-generation sequencing platforms do not cover the entire transcript length, the massive amount of data makes it possible to reconstruct transcripts in their full lengths.

5.2.1 Transcriptome Assembly Is Different from Genome Assembly

In the early times of ESTs, the same assemblers that were used for genomes were also used for transcriptomes. Although still technically possible, it is no longer the practice. There are fundamental differences between genome assembly and transcriptome assembly. In addition to the differences in uniformity of sequencing depth, the main difference is that in the genome assembly, the ideal output is a linear sequence representing each genomic region, whereas in the transcript assembly there can be several isoforms from the same locus, that is, the same exon of a gene is present in different contexts with other exons depending on the transcript. Therefore, in the transcript assembly, the gene is most naturally described as a graph, where nodes represent exons and arcs represent splicing events. Branches in node connections correspond to alternative splicing. An individual transcript is a single molecule which still should be represented as a linear sequence and it forms a path along the nodes in a graph. One exon node is present only once in an isoform, but the same exon can be in multiple different isoforms. The set of all possible paths in an exon graph includes all possible isoforms. The number of possible paths can be huge, but only a few of them are present in the real transcriptome. One of the challenges in the transcript assembly is to find which isoforms are real from all potential candidates. Once again, the problem comes from the short sequence read. If we could sequence an entire transcript at its full length in one read, the problem would be solved. The combinatorial problem emerges when we try to build long sequences from short fragments.

5.2.2 Complexity of Transcript Reconstruction

In order to illustrate the complexity of transcript reconstruction, let us take an example. If we suppose that there are three exons in a gene, the number of possible isoforms can be counted as a sum of the number of single exons, the number of exon pairs, and the number of exon triples. These numbers are 3, 3, and 1, which sum to 7. More generally, in the case of N exons, the number of possible isoforms is

$$\sum_{k=1}^{N} \binom{N}{k} = 2^N - 1$$

that is, there are two possibilities for each exon, either it is present in the isoform or it is not. The number 2^N also includes the case that the isoform is empty, that is, no exon is present, therefore one is subtracted from the total sum (corresponding to $k = 0$ in the equation above). This is just the number of possible isoforms. In the transcriptome, there can be any set of isoforms present. Although alternative splicing gives the possibility for many combinations, not all of them are present in a real transcriptome. The problem is to find which ones are true. The number of possible sets of isoforms is calculated along the same lines as above, one isoform is present or not present in the transcriptome, giving the number

$$2^{2^N - 1} - 1$$

which grows very fast as the number of exons N grows. For $N = 1, 2, 3, 4$, or 5, the sizes of possible isoform sets are 1, 7, 127, 32,767, and 2,147,483,647, respectively. This shows that already when the number of exons is greater than 4, it is not practical anymore to enumerate all possible solutions to test which one matches best with the data.

5.2.3 Assembly Process

There are two approaches for transcript reconstruction, mapping-based and *de novo* assembly. Both involve constructing a graph for each locus based on RNA-seq reads. The graph serves as a starting point for resolving isoforms. Both methods also include a problem of how to split the data so that a single graph represents only a single locus.

Mapping has been described in Chapter 4 of this book. Any method allowing split-reads can be used to align the RNA-seq reads on the genome. If gene models are available, this gives information about which exons belong to which genes. If no gene models are available, mapped reads must first be segmented in order to represent gene loci. An exon graph, also called a splicing graph, is then constructed for each locus and the task of finding a set of paths is applied within each graph, each path representing an isoform.

The number of possible isoforms can be reduced by limiting the connections in the exon graph. Each connection represents an exon junction. In a fully connected graph, all isoforms are possible since there is an arc between all nodes. The task is to choose a graph topology which best corresponds to the data. Those splicing events are removed for which there is no support from the RNA-seq reads, and only those connections in the graph are maintained which are needed. Evidence for maintaining an arc includes split-reads and paired-end information. In the case of a split-read, if the beginning of the read is mapped to one exon and the end of the read to another exon, this gives support for these two exons to be adjacent in a transcript sequence. In a paired-end case, this applies to the two ends of the read pair, one end is mapped to one exon and another end is mapped to another exon. Presence of a split-read is a stronger evidence for an exon junction than a paired-end read. In the case of mapped read pairs, insert's size information must be utilized to be sure that the two exons really form a junction, as opposed to the possibility that the two exons are merely in the same transcript but something else is between them. Insert size distribution depends on the RNA-seq library. Usually, the average insert size is used for each read pair and if the variance within the library is large, the estimate of an insert size for any particular read pair cannot be accurate.

In *de novo* assembly, there are basically two approaches: (1) to calculate pairwise overlaps between the reads which gives the topology of the assembly graph or (2) to construct a de Bruijn graph, which represents all sequence data as a set of k-mers and their connections. As a mathematical entity, the de Bruijn graph was introduced before the era of sequencing [3], and in the context of genome assembly it was first applied by Pevzner et al. [4]. The goal of *de novo* assembly is to extract as long as possible continuous segments (contigs) from the assembly graph which represent original parts of the genome or transcriptome. During the Human Genome Project in the 1990s, sequencing reads were relatively long (they were from Sanger sequencing) and their amount was less compared to today's data. Genome assemblers were based on a read-overlap approach, and the strategy was called overlap-layout-consensus (OLC) describing the three stages of the assembly. Although calculating all pairwise overlaps between reads is time-consuming, methodologically this is the easiest part of the problem. The main difficulty comes from the combinatorics: how to define the layout of the graph from which the consensus sequences of multiple read alignments are obtained. It is possible to construct an algorithm which finds the optimal solution for the assembly problem, but its execution time would be too long for any data set of practical value and therefore various heuristics and approximations must be used [5]. When the amount of sequence data increased, and at the same time reads became shorter, approaches utilizing de Bruijn graphs became more popular. In transcriptome assembly, most methods today are based on de Bruijn graphs. However, there are some exceptions, for example, MIRA EST assembler [6] which is based on an OLC paradigm.

5.2.4 de Bruijn Graph

Each node of a de Bruijn graph is associated with a $(k-1)$-mer. Two nodes A and B are connected if there is a k-mer whose prefix is the $(k-1)$-mer of the node A and the suffix is the $(k-1)$-mer of the node B. In this way, k-mers create edges in the de Bruijn graph [7]. Sequences are represented as paths in a graph, and even a single sequence read is spread to multiple connected nodes, the first node containing the $(k-1)$-mer starting from the first position of a sequence, the second node containing the $(k-1)$-mer starting from the second position of a sequence, and so on. Each k-mer is represented in the graph only once as an edge connecting two nodes. Two sequence reads share an edge if they have a common k-mer. This gives the information for the overlaps between reads and no pairwise comparisons need to be calculated explicitly. Construction of a de Bruijn graph is straightforward and much faster compared to calculating the overlaps between all read pairs. It consists of simply extracting all k-mers from reads and connecting the nodes representing the $(k-1)$-mers. The challenge then becomes how to find the paths in a graph which represent true transcripts. Sequencing errors result in tips which are dead ends in a graph and bubbles which complicate the structure of the graph. Bubbles are formed from branches in the graph which merge back together in another part of the graph. Some bubbles are due to sequencing error, but some are due to alternative splicing, for example, in the case of an exon in the middle of a gene model which is present in one isoform but skipped in another isoform. This results in two paths in the graph which share the beginning and end but have a branch in the middle. The k-mer order of single reads and paired-end read information are utilized when finding the paths in the graph. Edges can also be weighted by the abundance of k-mers which reduces erroneous paths. The length of the k-mer has an effect on the complexity of the graph. Clearly, it must be shorter than read length but if it is too small, the graph is dense in terms of connections since the nodes are not specific. However, if the k-mer is large, there must be enough data to make the graph connected. As a solution to the problem of choosing a suitable value for k, several assemblies can be done with different values of k, and the single best assembly is selected, or alternatively, the contigs from several assemblies with different values of k are combined [8–10].

5.2.5 Use of Abundance Information

If the set of candidate isoforms is of reasonable size, it is possible to use RNA-seq abundance information to solve the isoforms. The reasoning is that the abundance should be the same in all exons belonging to the same transcript. One transcript is one molecule, so if there are no biases in library preparation and sequencing (and mapping), sequence reads should cover and represent an entire transcript uniformly. If there are deviations from this, for example, some exons have larger sequence depth, then it indicates that those exons are also present in other isoforms.

For a fixed set of isoforms, it is possible to estimate their relative abundances. The optimization task is to get abundances which best describe the data. This can be done by first setting initial values for abundances, for example, by dividing the abundances evenly among all isoforms, and then fine-tuning the solution iteratively using an Expectation Maximum (EM) algorithm. The EM algorithm was introduced in the 1970s [11], in the context of transcriptome data it is described in [12]. The optimization consists of iterations of two steps: Expectation (E) and Maximization (M). In the E-step, all reads are assigned proportionally to each isoform according to the isoform abundance, and in the M-step, relative abundances of isoforms are recalculated. These two steps are repeated until the estimated abundance values no longer change, that is, the algorithm has converged. The solution is for the given set of isoforms, so if new isoforms are added to the set of existing isoforms, all values may change. In general, since the EM algorithm finds a local optimum, in case there are multiple local optima, the solution depends on the initial values. However, in the case of a linear model with nonnegative parameters, there is only one maximum, so the local optimum is also the global optimum [13]. Basic EM algorithm can be modified in many ways. For example, in iReckon software [14], the regularized EM algorithm is used in order to reduce the number of spurious transcript reconstructions.

5.3 DATA PREPROCESSING

Typically, base call quality diminishes toward the end of a read. This is characteristic to first- and second generation sequencing technologies (Sanger, Illumina, SOLiD, 454) but does not necessarily apply to new sequencing technologies (such as PacBio). If alignment quality is calculated along the entire read length, the low-quality part of a read with more errors reduces the alignment score. Therefore, by trimming the low-quality part of the read, the number of mappable reads can be increased. Also, in *de novo* assembly, if it is based on pairwise read overlaps, it is beneficial to trim low-quality parts of reads. In de Bruijn graph-based methods, however, erroneous tails of the reads result in tips and dead ends in the graph, but because the graph is based on *k*-mers, the low-quality end of the read does not affect the *k*-mers in the beginning of the read. Trimming of reads simplifies the graph and reduces the number of dead ends, but including erroneous and low-quality parts of the reads does not completely prevent the assembly. However, an excessive amount of low-quality data may affect assembly and a massive amount of data will in any case slow down the computation. Erroneous reads increase the number of nodes in a de Bruijn graph and therefore increase the memory use. Nevertheless, it might be a good idea to first do the assembly with the data as it is. This gives a way to compare the results and see the effect of trimming. And most importantly, it is a good way to check if something went wrong in the trimming process.

Artifacts caused by the library construction should be removed from reads regardless of the assembly method. These artifacts include adaptor sequences which might be remaining in a portion of the sequence reads. Also, if polyA is included in sequencing, it should be trimmed off.

The user should know how the sequencing library was constructed and how the reads are oriented. In Illumina paired-end reads, the reads face each other. Another information is the strand specificity of the library. It is possible to construct sequence libraries so that the strand where the reads are from is known. Strand-specific libraries give an advantage to resolve overlapping genes which are in opposite strands.

5.3.1 Read Error Correction

Read filtering and trimming are means to get rid of sequencing errors by removing entire reads or parts of them. These procedures reduce the amount of sequence data. A completely different idea is to try to correct the errors in the reads. If this is successful, there are more useful data available.

One of the main applications of read correction is *de novo* assembly. Using de Bruijn graph-based assemblers, each k-mer (actually $(k - 1)$-mer) allocates a node in the graph. Sequencing errors result in a number of incorrect k-mers and produce useless nodes which both slow down the computation and increase the memory use. However, not all variation in the data are random due to sequencing errors, in diploid and polyploid organisms; there can be non-random variation due to differences between alleles. In some cases, it might be beneficial to do "overcorrection" and also eliminate these kinds of variations. If SNPs and indels are removed from sequence reads, data become more homogeneous, de Bruijn graph is simplified, and longer contigs can be produced. Sequence variants can later be detected by mapping original uncorrected reads against the contigs.

Read correction is based on utilizing redundancy in the data. In order to work properly, there must be enough sequencing depth. If reads were aligned perfectly without alignment errors against the genome or transcriptome, it would be easy to detect sequencing errors and correct them by majority voting. The challenge comes when there is no reference available and there are similar sequences originating from different parts of genome due to repeats or otherwise similar regions.

5.3.2 Seecer

The first error correction software purposely dedicated to RNA-seq data is SEECER [15]. It works by correcting reads one by one. For each read to be corrected, other reads are selected which share at least one k-mer with it. Clustering is applied in order to separate reads coming from different transcripts. A subset of reads is used for building a hidden Markov model (HMM), a probabilistic model to represent the group of sequences. Reads are then aligned against the states of the HMM using the Viterbi algorithm, and read correction is based on the consensus of the HMM. All those sequences whose likelihood exceeds a given threshold, an indication that they match well enough with the model, are corrected. Once the read is corrected, it is removed from the pool of available sequences for correction and the process is repeated for the remaining data.

Error correction requires memory and it might not work with standard desktop computer. There should be some tens of gigabytes RAM available, depending on the size of data and read length. SEECER can be downloaded from http://sb.cs.cmu.edu/seecer/. The steps required for error correction are implemented in Bash shell script run_seecer.sh which takes input files in FASTA or FASTQ format. *k*-mers can be calculated using internal implementation or external software Jellyfish. Latter choice is recommended especially with large data sets. Default *k*-mer length is 17.

Running SEECER requires GNU Scientific Library (GSL). In order to install it in default location requires sudo rights.

1. Get gsl-1.16.tar.gz from http://ftpmirror.gnu.org/gsl/ (in our case, closest ftp mirror was http://www.nic.funet.fi/pub/gnu/ftp.gnu.org/pub/gnu/gsl/)

```
$ tar xvfz gsl-1.16.tar.gz

$./configure

$ make

$ sudo make install
```

2. Get SEECER-0.1.2.tar.gz from http://sb.cs.cmu.edu/seecer/install.html

```
$./configure

$ make
```

3. Run SEECER. Options can be listed with –h parameter

```
$ bash bin/run _ seecer.sh -h
```

Create temporary directory "tmp" for computation and run read correction. Files "reads1.fq" and "reads2.fq" contain paired-end reads.

```
$ mkdir tmp

$ bash bin/run_seecer.sh -t tmp reads1.fq reads2.fq
```

Corrected reads are in FASTA format with suffix "_corrected.fa" in the same directory with original reads.

5.4 MAPPING-BASED ASSEMBLY

Here we describe two software packages, Cufflinks and Scripture, which are used to reconstruct full-length transcript sequences based on RNA-seq read mapping. Both can be used for *ab initio* reconstruction of transcripts, that is, there is no need to have external gene models. The main difference between these two programs is the methodology to resolve isoforms: Scripture reports all possible isoforms, whereas Cufflinks reports the smallest possible set of isoforms which can explain the data. Output is given in BED or GTF format which contains the transcript coordinates in a reference sequence. Since reference sequence is known, it is straightforward to convert the transcript sequence coordinates into FASTA file using any scripting language, for example, Python or Perl.

Mapping can be done with TopHat, and version 2.0 is used here. Input data consist of a FASTA file of chromosome 18 "chr18.fa" and paired-end read files "chr18_1.fq" and "chr18_2.fq." Burrows–Wheeler transformed index files are named "chr18" and mapping output will be in directory "top2." Since reads are 2 × 75 bp and fragment insert size 200 bp, the inner distance between reads is 50 bp which is given as a parameter with the argument "–r" in TopHat. Here, mapping is done with four threads. In order to use TopHat, both SAMtools and Bowtie (here Bowtie2) must be available so that their locations must be included in the PATH variable.

```
$ bowtie2-build chr18.fa chr18

$ tophat2 -r 50 -p 4 -o top2 chr18 chr18_1.fq chr18_2.fq
```

5.4.1 Cufflinks

Cufflinks is written in C++ [13]. It has been updated actively and the most recent version can be downloaded from http://cuffflinks.cbcb.umd.edu. The web page contains a user manual and further information. Cufflinks works by first constructing a graph which parsimoniously explains the data. That is, it finds the smallest set of transcripts which is able to represent the RNA-seq reads. Abundances are then estimated for this set of transcripts.

At the time of writing this book, version 2.1.1 is the most recent and it supports BAM files generated by TopHat2. In order to utilize paired-end information, read names in BAM files should not contain read pair suffices. Although TopHat correctly indicates paired-end information in the BAM file, that is, flag " = " is present if both ends have been mapped, it does not remove paired end-read suffices if they are not separated by expected delimiters. For example, suffices "/1" and "/2" are automatically removed from read names, whereas suffices "_1" and "_2" are not. In order for Cufflinks to utilize paired-end information, both reads of the read pair should have the same identifier in the BAM file (first column in the file). This can easily be checked using SAMtools view command with the BAM file as input.

In order to use Cufflinks, the location of SAMtools must be in the PATH variable.

There are several parameters which can be defined. In order to speed up the computation, four threads are used in the command below; otherwise, default parameters are used and the command for running Cuffinks is

```
$ cufflinks -p 4 -o outdir top2/accepted _ hits.bam
```

Gene models are stored in a GTF file in the output directory. There are four output files

```
-rw------- 1   somervuo   50K    Jul   15 10:43   genes.fpkm_tracking
-rw------- 1   somervuo   67K    Jul   15 10:43   isoforms.fpkm_tracking
-rw------- 1   somervuo   0      Jul   15 10:42   skipped.gtf
-rw------- 1   somervuo   898K   Jul   15 10:43   transcripts.gtf
```

Transcripts with the exon information are in file "transcripts.gtf." In this case, there are 750 transcripts from 634 genes. These are listed in files "isoforms.fpkm_tracking" and "genes.fpkm_tracking," respectively.

If there are several libraries with different insert sizes, it is better to run Cufflinks separately to each of them and then merge the results, rather than concatenating first all BAM files and then running Cufflinks. The program Cuffmerge can be used for merging several Cufflinks runs.

The amount of merging isolated segments can be controlled with the argument "`--overlap-radius`." The default value is 50 in base pairs. Larger value results in merging of more distant gene models.

In the example above, no existing knowledge of gene models was utilized. If there is such information, it can be utilized by giving a GTF file as a guidance to Cufflinks using the argument "–g."

For the comparison of Cufflinks output with existing gene models, there is a program Cuffcompare. If reference gene models are in file "ref. gtf," the command is

```
$ cuffcompare -r ref.gtf transcripts.gtf
```

Output files contain the summary and gene-wise information for the similarity between the gene models in the two files.

5.4.2 Scripture

Scripture is a Java-based software [16]. It can be downloaded from http://www.broadinstitute.org/software/scripture/. Scripture segments the data based on split read information. Regions of a genome with split-read connections form islands which can further be connected using paired-end read information. Isoforms within these regions are reported.

Scripture starts by constructing a connectivity graph. It contains all bases of a reference genome as its nodes. Two nodes are connected if the corresponding two bases are adjacent in a genome or in a transcript. Split-reads give information for exon–intron borders, and each connection must be supported by at least two RNA-seq reads. The allowed donor/acceptor splice sites are canonical GT/AG and noncanonical GC/AG and AT/AC. The paths in the connectivity graph are evaluated for their statistical significance of how much they are enriched compared to the background read mapping distribution. This is implemented by scanning the graph with fixed-sized windows and assigning a *p*-value to each window. Significant windows are merged to create a transcript graph which are refined using paired-end reads to link previously isolated segments.

The input data for Scripture is a sorted BAM file and a reference chromosome FASTA file. There is a new version 2.0 of Scripture which was not publicly available at the time of writing the book, but its preliminary version was obtained by its authors. The syntax is

```
$ java -jar ScriptureVersion2.0.jar -task reconstruct
  -alignment top2/accepted_hits.bam -genome chr18.fa
  -out out -strand unstranded -chr 18
```

The output of the earlier version of Scripture consisted of two files, one containing the gene models in BED format and another file containing the transcript graphs in DOT format. In version 2.0, there are four output files. In addition, the new version creates a coordinate file in the same directory where the BAM file is located. The four output files are

```
-rw------- 1 somervuo 80K Jul 8 15:13 out.connected.bed
-rw------- 1 somervuo 250K Jul 8 14:09 out.pairedCounts.txt
```

```
-rw------- 1 somervuo 229K Jul 8 14:09 out.pairedGenes.bed
-rw------- 1 somervuo 104K Jul 8 14:09 out.scripture.paths.bed
```

File “out.scripture.paths.bed” reports the initial transcripts utilizing only single read information, and “out.connected.bed” reports the transcripts where paired-end information has been utilized. In the latter file, there are 549 transcripts from 504 genes.

5.5 *DE NOVO* ASSEMBLY

Here we describe two software packages which are used to reconstruct full-length transcript sequences *de novo*, that is, without the help of a reference genome. Both of them utilized de Bruijn graphs. The first one consists of two programs, Velvet and Oases. Velvet is a genome assembler which produces one assembly graph that is used by the second program Oases to find paths which represent isoforms. The other assembly program is Trinity which consists of three modules. First, the RNA-seq reads are initially assembled and clustered, each cluster representing a locus in a genome. A de Bruijn graph is constructed for each cluster, and linear transcript sequences are extracted so that there can be several isoforms from the same locus. Both software tools copy the sequence data into one file before the assembly, so if large data sets are used, disk space should be checked before starting the assemblies.

5.5.1 Velvet + Oases

Velvet is written in C. It was introduced as a genome assembler [17]. Later, another program Oases was written for transcript assembly which utilizes the output of Velvet [9]. Velvet can be downloaded from http://www.ebi.ac.uk/~zerbino/velvet/ and Oases from http://www.ebi.ac.uk/~zerbino/oases/. Both software packages include well-written manuals.

Velvet consists of two programs: velveth and velvetg. The first one calculates *k*-mers of data and the second one finds and extracts contigs in a de Bruijn graph. Oases segments the graph and extracts isoforms from each locus. Transcript sequences obtained by Oases are usually much longer compared to the contigs of Velvet. In order to use paired-end reads in Velvet, they must be interleaved, that is, both reads of a read pair are located adjacently in the same file. There is a Perl script in the Velvet package to do interleaving if read pairs are stored originally in two separate files. This command creates a new file “chr18_12.fq” where the reads are interleaved

```
$ shuffleSequences_fastq.pl chr18_1.fq chr18_2.fq chr18_12.fq
```

The first task is to create a hash table: here we define *k*-mer length to be 25 and the output directory will be "vdir." The data format is also defined; in this case, paired-end reads in FASTQ format. Graph traversal and contig extraction is done in the second step. Here we define insert size to be 200 bp. In Velvet, insert size is the fragment length, that is, it includes the read lengths. It is important to include the argument "-read_trkg" with parameter "yes," since Oases utilizes read tracking information.

```
$ velveth vdir 25 -fastq -shortPaired chr18_12.fq
$ velvetg vdir -ins_length 200 -read_trkg yes
```

Oases is applied to the resulting de Bruijn graph. The input for Oases is the name of the directory which contains the Velvet output. In the case of paired-end reads, insert size must also be defined. Here, the minimum transcript length is defined to be 200 bp.

```
$ oases vdir -ins_length 200 -min_trans_lgth 200
```

The output directory vdir contains the files shown below. Transcript sequences are stored in a FASTA file "transcripts.fa." The name of each FASTA entry describes the locus and isoform. Another file produced by Oases is "contig-ordering.txt."

```
-rw------- 1  somervuo   25M  Jul 16 11:56  Graph2
-rw------- 1  somervuo   11M  Jul 16 11:59  LastGraph
-rw------- 1  somervuo  1.2K  Jul 16 11:59  Log
-rw------- 1  somervuo  5.5M  Jul 16 11:56  PreGraph
-rw------- 1  somervuo   34M  Jul 16 11:55  Roadmaps
-rw------- 1  somervuo   84M  Jul 16 11:55  Sequences
-rw------- 1  somervuo  1.3M  Jul 16 11:59  contig-ordering.txt
-rw------- 1  somervuo  2.6M  Jul 16 11:56  contigs.fa
-rw------- 1  somervuo  253K  Jul 16 11:59  stats.txt
-rw------- 1  somervuo  1.6M  Jul 16 11:59  transcripts.fa
```

An example of a FASTA entry name is "Locus_10_Transcript_1/3_Confidence_0.571_Length_3815." It indicates that there are three transcripts from locus 10, and this is the first of them. The confidence value is a number between 0 and 1 (the higher the better), and length is the transcript length in base pairs. In this example, in file "transcripts.fa" there are 1308 transcript sequences with a minimum length of 200 bp from 862 loci.

In Oases version 0.2, it is possible to run several assemblies with different k-mer lengths and merge the assemblies. In the Oases package, there is a Python script for this purpose. Here we define that all odd k-mers from 19 to 29 will be used. Additional parameters for Velvet and Oases are given with arguments "–d" and "–p." Using the Python script, there is no need to use the argument "–read_trkg."

```
$ python oases_pipeline.py -m 19 -M 29 -o odir -d " -fastq -
  shortPaired chr18_12.fq" -p " -ins_length 200 -min_trans_
  lgth 200"
```

This produces separate output for each k-mer and one directory "odir-Merged" which contains the results of a merged assembly. In this example, file "transcripts.fa" in odirMerged contains 4468 transcript sequences from 827 loci. After creating output directories for each k-mer, it is possible to merge only some of the assemblies without starting from the beginning. This is done by using the argument "–r" in Python script. For example, if only the assemblies with $k = 25$ and larger are merged, which is done by

```
$ python oases_pipeline.py -m 25 -M 29 -r -o odir
```

It produces 2159 transcripts from 783 loci.

By default, the maximum k-mer length in Velvet is 31; however, it is possible to use also larger values. For example, if values of k up to 51 are needed, Velvet can be compiled with the command (and the same must be applied for compiling Oases)

```
$ make 'MAXKMERLENGTH=51'
```

When running several assemblies with different values for k, it is beneficial to use arguments "–m" and "–M" rather than launching each assembly separately since velvet copies the read data into file Sequences, but with "–m" and "–M" parameters, it does the copying only once and the other directories contain symbolic links to the file in the first output directory.

5.5.2 Trinity

Trinity software package [18] can be downloaded from http://trinityrnaseq.sourceforge.net/. The web page contains lots of useful information describing the method and it also includes advanced topics. Trinity-based workflow including downstream analysis is also described in [19].

Trinity consists of three separate programs: (1) Inchworm, which constructs initial contigs, (2) Chrysalis, which clusters the contigs produced by Inchworm and creates a de Bruijn graph for each locus, and (3) Butterfly, which extracts the isoforms within each de Bruijn graph. The word "component" is used instead of locus in Trinity. During the Butterfly step, it is possible that a component produced by Chrysalis will be divided into smaller pieces if it seems that sequence reads come from more than one locus. If this happens, it is reported in the names of the output transcript sequences.

All three programs can be run using one Perl script Trinity.pl. Below we define that our sequences are in FASTQ format and the number of processors for computation is 4. *k*-mer length is fixed to be 25 in the current versions of Trinity. Jellyfish, software used for calculating *k*-mers, requires defining maximum memory and in this case it is set to 10G. The shortest transcript to be reported is 200 bp by default. Before running Trinity, stack size should be defined to be unlimited. This can be done using the shell command `unlimit` or `ulimit -s unlimited`, depending on the Linux distribution. There are some differences between different versions of Trinity, for example, in earlier versions the user needed to define which *k*-mer method to use and there are also differences in the number of output files. At the time of writing the book, the most recent version is r2013-02-25 where the Jellyfish method is used by default. The command to run Trinity with default parameters and 4 cpus is

```
$ Trinity.pl --seqType fq --JM 10G --left chr18_1.fq
  --right chr18_2.fq --CPU 4
```

If there is no name defined for output directory, it will be "trinity_out_dir." After the Butterfly process has finished, the output directory contains a FASTA file "Trinity.fasta" which contains all isoforms. It is possible that one or more graphs in a Butterfly step do not produce any transcript sequences. However, all information is stored in the output directory under the subdirectory "chrysalis." There is a separate FASTA file for each component which contains its transcript sequences. The graph structure of the component is also stored. If Butterfly failed to generate any transcript sequence for the component, the corresponding FASTA file exists but its size is zero. Using the example data, the output directory looks like the following:

```
-rw------- 1 somervuo   2.2M  Dec 18 15:13 Trinity.fasta
-rw------- 1 somervuo    583  Dec 18 15:13 Trinity.timing
-rw------- 1 somervuo    78M  Dec 18 14:56 both.fa
-rw------- 1 somervuo      7  Dec 18 14:56 both.fa.read_count
-rw------- 1 somervuo   159M  Dec 18 14:59 bowtie.nameSorted.sam
-rw------- 1 somervuo      0  Dec 18 14:59 bowtie.nameSorted.sam.
   finished
-rw------- 1 somervuo      0  Dec 18 14:59 bowtie.out.finished
drwx------- 3 somervuo  4.0K  Dec 18 15:04 chrysalis
-rw------- 1 somervuo   3.6M  Dec 18 14:58 inchworm.K25.L25.DS.fa
-rw------- 1 somervuo      0  Dec 18 14:58 inchworm.K25.L25.DS.fa.
   finished
-rw------- 1 somervuo      8  Dec 18 14:58 inchworm.kmer_count
-rw------- 1 somervuo   148K  Dec 18 14:59 iworm_scaffolds.txt
-rw------- 1 somervuo      0  Dec 18 14:59 iworm_scaffolds.txt.
   finished
-rw------- 1 somervuo      0  Dec 18 14:57 jellyfish.1.finished
-rw------- 1 somervuo   125M  Dec 18 14:57 jellyfish.kmers.fa
-rw------- 1 somervuo    13M  Dec 18 14:59 scaffolding_entries.sam
-rw------- 1 somervuo   6.3M  Dec 18 14:59 target.1.ebwt
-rw------- 1 somervuo   279K  Dec 18 14:59 target.2.ebwt
-rw------- 1 somervuo   170K  Dec 18 14:58 target.3.ebwt
-rw------- 1 somervuo   557K  Dec 18 14:58 target.4.ebwt
lrwxrwxrwx 1 somervuo     73  Dec 18 14:58 target.fa ->/.../
   inchworm.K25.L25.DS.fa
-rw------- 1 somervuo      0  Dec 18 14:59 target.fa.finished
-rw------- 1 somervuo   6.3M  Dec 18 14:59 target.rev.1.ebwt
-rw------- 1 somervuo   279K  Dec 18 14:59 target.rev.2.ebwt
```

In this example, file "Trinity.fasta" contains 1837 transcripts from 1293 components. For comparison, when minimum *k*-mer coverage was increased from 1 (which is default) to 2 using argument "—min_kmer_cov 2" in the command line, the resulting assembly contained 1205 transcripts in 848 components. In the output directory, file "both.fa" contains all input sequence data. The two paired-end read files without quality information are concatenated there. Directory "chrysalis" contains the following files:

```
drwx------- 4 somervuo 4.0K Dec 18 15:02 Component_bins
-rw------- 1 somervuo     0 Dec 18 15:00 GraphFromIwormFasta.finished
-rw------- 1 somervuo  2.1M Dec 18 15:00 GraphFromIworm Fasta.out
-rw------- 1 somervuo  1.5M Dec 18 15:00 bundled_iworm_contigs.fasta
-rw------- 1 somervuo   57M Dec 18 15:02 bundled_iworm_contigs.fasta.
   deBruijn
-rw------- 1 somervuo     0 Dec 18 15:00 bundled_iworm_contigs.fasta.
   finished
-rw------- 1 somervuo  507K Dec 18 15:03 butterfly_commands
-rw------- 1 somervuo  507K Dec 18 15:13 butterfly_commands.completed
-rw------- 1 somervuo     0 Dec 18 15:02 chrysalis.finished
```

```
-rw------- 1 somervuo  138K  Dec 18 15:03 component_base_listing.txt
-rw------- 1 somervuo     0  Dec 18 15:03 file_partitioning.ok
-rw------- 1 somervuo  643 K Dec 18 15:03 quantifyGraph_commands
-rw------- 1 somervuo  643 K Dec 18 15:04 quantifyGraph_commands.
  completed
-rw------- 1 somervuo     0  Dec 18 15:04 quantifyGraph_commands.run.
  finished
-rw------- 1 somervuo     7  Dec 18 15:02 rcts.out
-rw------- 1 somervuo     0  Dec 18 15:02 readsToComponents.finished
-rw------- 1 somervuo   79M  Dec 18 15:02 readsToComponents.out.sort
-rw------- 1 somervuo     0  Dec 18 15:02 readsToComponents.out.sort.
  finished
```

Directory "Component_bins" contains subdirectories "Cbin0," "Cbin1," etc., the number depending on the total number of components resulting from assembly. An example of a component which produced more than one transcript is shown below. Files with suffix ".dot" are for visualizing the graphs. In the current Trinity version r2013-02-25, they are not produced by default, but they can be generated by providing an argument "—bfly_opts '-V 5'" in the command line when running Trinity.

```
-rw------- 1 somervuo    5.9K   Dec 18 16:49 c420.graph.allProb
  Paths.fasta
-rw------- 1 somervuo    134K   Dec 18 16:23 c420.graph.out
-rw------- 1 somervuo    1.8M   Dec 18 16:23 c420.graph.reads
-rw------- 1 somervuo     492   Dec 18 16:49 c420.graph_final
  CompsWOloops.L.dot
-rw------- 1 somervuo     492   Dec 18 16:49 c420.graph_withLoops.
  J.dot
```

There are three transcript sequences in file "c420.graph.allProbPaths. fasta," and their names are

```
>c420.graph_c0_seq1 len=  328   path=[305894:0-327]
>c420.graph_c1_seq1 len=  2675  path=[287873:0-149 288298:150-2674]
>c420.graph_c1_seq2 len=  2730  path=[287873:0-149 288178:150-204
  288298:205-2729]
```

There are two subcomponents c0 and c1, which means that the original component c420 was divided into two during the Butterfly process. The name also includes the length of the transcript sequence and the node path in a de Bruijn graph. In Figure 5.1, the graph file "c420.graph_withLoops.J.dot" is shown using the program GraphViz. The longer transcript (seq2) contains a 55-bp segment which is missing in the shorter transcript (seq1) in the component c1. A 328-bp transcript corresponding to c0 is located in the top right corner.

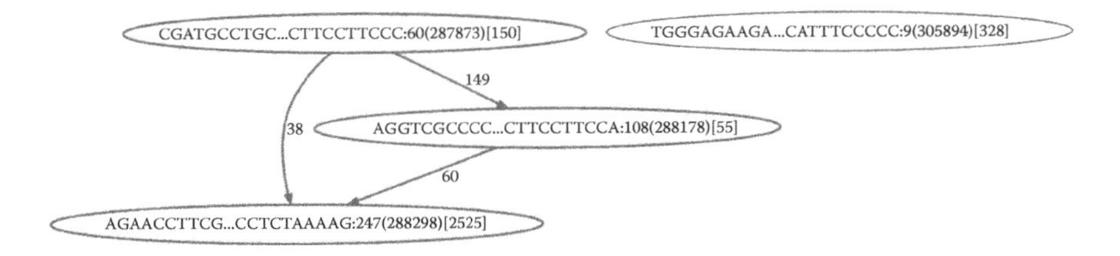

FIGURE 5.1 Example transcript graph resulting from Trinity assembly.

In order to compare the longest transcript of c420 against the known transcripts, it was mapped against the human genome using the UCSC genome browser (http://genome.ucsc.edu/). The best BLAT hit came from chromosome 18. In Figure 5.2, Trinity transcript sequence is shown on the top labeled as "YourSeq."

ASSEMBLING TRANSCRIPTS IN CHIPSTER

The tool category "RNA-seq" offers currently the Cufflinks package for transcriptome assembly.

- Select the BAM file you produced with TopHat2 in Chapter 4, and the tool "Assemble reads into transcripts using Cufflinks." Note that you can also give a GTF file as input if you would like to use existing annotation to guide the assembly. Check that the files were assigned correctly by scrolling to the end of the parameter panel, and click "Run." Note that the parameters allow you to also correct the abundance estimates for multimapping reads and sequence-specific bias.
- You can visualize the gene models by opening the output file "transcripts.gtf" in Chipster genome browser as described in Chapter 4. In order to navigate efficiently from one transcript to another, open the GTF file in a separate window and click on the start coordinate.
- You can merge assemblies from several samples using the Cuffmerge tool, and you can compare assemblies using the Cuffcompare tool.

5.6 SUMMARY

This chapter has described the basic methodology and four software packages for reconstructing transcript sequences based on short RNA-seq reads. Numbers of loci and transcripts resulting from each assembly with sample data have been reported for the purpose of reproducibility. They are not meant to be used as such for comparing the methods. Comparison of several tools for transcript reconstruction, including Cufflinks and

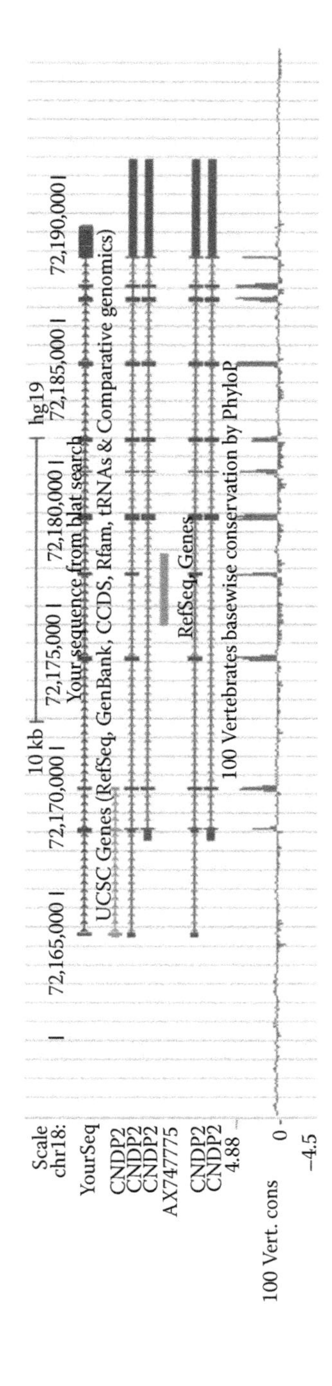

FIGURE 5.2 Assembled transcript "YourSeq" mapped on human genome and shown in UCSC genome browser.

Oases, can be found in [20]. Trinity and Oases-based *de novo* assemblies have been analyzed in [21]; see also [9] which includes Cufflinks in the comparison.

Besides the data and preprocessing, output of an assembly depends on the parameter setting particular to each software. The number of transcripts can be easily varied by changing the parameters related to minimum contig length and coverage. But the plain number and length of transcripts reveal nothing about the accuracy and errors of assembly. In fact, measuring the quality of an assembly is not a straightforward task. Especially it is difficult when there are no references or previously known gene models available. In practice, there is a tradeoff between specificity and sensitivity. Stringency should reduce errors but if there is not enough read coverage, contigs become short and originally long transcripts become fragmented. Although both mapping and *de novo*-based assembly methods have been shown to be able to reconstruct full-length transcripts from short sequence reads, it is important to be aware that this depends on the data quality and coverage. Many challenges of transcript assembly will disappear when the technology will be mature to allow sequencing full-length transcript in one read. At the moment, such reads come from a third-generation PacBio sequencer. However, although it provides long reads, the drawback is the limited sequencing depth per sequencing run which makes other platforms more cost efficient at the moment. For this reason, second-generation sequencing and the methods described in this chapter continue to be used for now and at the same time new tools are also likely to be developed.

REFERENCES

1. Adams M.D., Kelley J.M., Gocayne J.D., et al. Complementary DNA sequencing: Expressed sequence tags and human genome project. *Science* 252(5013):1651–1656, 1991.
2. Quackenbush J., Liang F., Holt I., Pertea G., and Upton J. The TIGR gene indices: Reconstruction and representation of expressed gene sequences. *Nucleic Acids Research* 28(1):141–145, 2000.
3. de Bruijn N.G. A combinatorial problem. *Koninklijke Nederlandse Akademie v. Wetenschappen* 49:758–764, 1946.
4. Pevzner P., Tang H., and Waterman M. An Eulerian path approach to DNA fragment assembly. *Proceedings of the National Academy of Sciences of United States of the America* 98(17):9748–9753, 2001.
5. Kececioglu J. and Myers E. Combinatorial algorithms for DNA sequence assembly. *Algorithmica* 13:7–51, 1995.

6. Chevreux B., Pfisterer T., Drescher B., et al. Using the miraEST assembler for reliable and automated mRNA transcript assembly and SNP detection in sequenced ESTs. *Genome Research* 14:1147–1159, 2004.
7. Compeau P., Pevzner P., and Tesler G. How to apply de Bruijn graphs to genome assembly. *Nature Biotechnology* 29(11):987–991, 2011.
8. Robertson G., Schein J., Chiu R., et al. *De novo* assembly and analysis of RNA sequence data. *Nature Methods* 7(11):909–912, 2010.
9. Schulz M.H., Zerbino D.R., Vingron M., and Birney E. Oases: Robust *de novo* RNA-seq assembly across the dynamic range of expression levels. *Bioinformatics* 28(8):1086–1092, 2012.
10. Surget-Groba Y. and Montoya-Burgos J. Optimization of *de novo* transcriptome assembly from next-generation sequencing data. *Genome Research* 20(10):1432–1440, 2010.
11. Dempster A.P., Laird N.M., and Rubin D.B. Maximum likelihood from incomplete data via the EM algorithm. *Journal of the Royal Statistical Society. Series B (Methodological)* 39(1):1–38, 1977.
12. Xing Y., Yu T., Wu Y.N., Roy M., Kim J., and Lee, C. An expectation-maximization algorithm for probabilistic reconstructions of full-length isoforms from splice graphs. *Nucleic Acids Research* 34(10):3150–3160, 2006.
13. Trapnell C., Williams B.A., Pertea G., et al. Transcript assembly and quantification by RNA-Seq reveals unannotated transcripts and isoform switching during cell differentiation. *Nature Biotechnology* 28(5):511–515, 2010.
14. Mezlini A., Smith E., Fiume M., et al. iReckon: Simultaneous isoform discovery and abundance estimation from RNA-seq data. *Genome Research* 23: 519–529, 2013.
15. Le H., Schulz M., McCauley B., Hinman V., and Bar-Joseph Z. Probabilistic error correction for RNA sequencing. *Nucleic Acids Research* 41(10):e109, 2013.
16. Guttman M., Garber M., Levin J.Z., et al. *Ab initio* reconstruction of cell type-specific transcriptomes in mouse reveals the conserved multi-exonic structure of lincRNAs. *Nature Biotechnology* 28(5):503–510, 2010.
17. Zerbino D.R. and Birney E. Velvet: Algorithms for *de novo* short read assembly using de Bruijn graphs. *Genome Research* 18(5):821–829, 2008.
18. Grabherr M.G., Haas B.J., Yassour M., et al. Full-length transcriptome assembly from RNA-seq data without a reference genome. *Nature Biotechnology* 29(7):644–652, 2011.
19. Haas B.J., Papanicolaou A., Yassour M., et al. *De novo* transcript sequence reconstruction from RNA-seq using the Trinity platform for reference generation and analysis. *Nature Protocols* 8(8):1494–1512, 2013.
20. Steijger T., Abril J.F., Engström P.G., et al. Assessment of transcript reconstruction methods for RNA-seq. *Nature Methods* 10(12):1177–1184, 2013.
21. Francis W.R., Christianson L.M., Kiko R., Powers M.L., Shaner N.C., and Haddock S.H. A comparison across non-model animals suggests an optimal sequencing depth for *de novo* transcriptome assembly. *BMC Genomics* 14:167, 2013.

CHAPTER 6

Quantitation and Annotation-Based Quality Control

6.1 INTRODUCTION

Once reads have been mapped to a reference genome, their location can be matched with genomic annotation. This enables us to quantitate gene expression by counting reads per genes, transcripts and exons, and it also opens up new possibilities for quality control. Quality aspects that can be measured only with mapped reads include saturation of sequencing depth, read distribution between different genomic feature types, and coverage uniformity along transcripts. The first half of this chapter introduces these annotation-based quality metrics and presents some software for checking them.

The second half discusses quantitation of gene expression, which is an integral part of most RNA-seq studies. In principle, calculating the number of mapped reads provides a direct way to estimate transcript abundance, but in practice several complications need to be taken into account. Eukaryotic genes typically produce several transcript isoforms via alternative splicing and promoter usage. However, quantitation at transcript level is not trivial with short reads, because transcript isoforms often have common or overlapping exons. Furthermore, the coverage along transcripts is not uniform because of mappability issues and biases introduced in library preparation. Because of these complications, expression is often estimated at the gene level or the exon level instead. However, gene level counts are not optimal for differential expression analysis for those genes which undergo isoform switching, because the number of counts depends on transcript length. This issue is described in more detail in Chapter 8 in the context of differential expression analysis.

6.2 ANNOTATION-BASED QUALITY METRICS

As discussed in Chapter 3, laboratory protocols for producing RNA-seq data are not perfect yet, but luckily many read quality problems such as low-confidence bases and biases in nucleotide composition can be detected already at the raw read level. However, some important quality aspects can be measured only when reads have been mapped to a reference genome and their location is matched with annotation. These include the following:

- *Saturation of sequencing depth.* The reliability of expression profiling, splicing analysis, and transcript construction depends on sequencing depth. Because sequencing is costly, it is important to check how close to saturation the data are, that is, would new genes and splice junctions be discovered with additional sequencing. Ideally the right depth would be determined beforehand of course, but this would require a data set from the same species and tissue, because saturation depends on transcriptome complexity.

- *Read distribution between different genomic features.* This can be done at several levels, for example, reads can be counted in exonic, intronic, and intergenic regions, and exonic reads can be further distributed between coding, 5′UTR and 3′UTR exons. If a high proportion of reads map to intronic or intergenic regions, it might be worth looking for novel isoforms and genes, but this could also be a sign of contaminating genomic DNA. Reads mapping to genes can further be distributed to biotypes such as protein coding genes, pseudogenes, ribosomal RNA (rRNA), miRNAs, etc. The rRNA content is particularly important, because laboratory protocols for removing rRNA can be unreliable and inconsistent between samples. If a large fraction of your reads map to rRNA, you can remove them, for example, by mapping them against rRNA sequences with Bowtie2 (as described in Chapter 4) and keeping the unaligned reads.

- *Coverage uniformity along transcripts.* Different laboratory protocols can introduce different location biases. For example, protocols which include a poly-A capture step can result in reads which are predominantly from the 3′ ends of transcripts. This 3′ bias can vary between samples, so it is important to estimate the degree of it.

6.2.1 Tools for Annotation-Based Quality Control

Several quality control tools for aligned RNA-seq data are available, including RSeQC [1], RNA-SeQC [2], Qualimap [3], and Picard's CollectRNASeqMetrics tool [4]. They report many overlapping quality measures, but have also their individual strengths. All of them offer command line interface, RNAseQC and Qualimap have also their own GUI, and RseQC is available in the Chipster software. The annotation information is typically given in GTF [5] or BED [6] files, which need to have the same chromosome naming as BAM files.

RNA-SeQC is implemented in Java and it takes annotation in GTF format. It also requires a reference FASTA file with an index (.fai) and a sequence dictionary file (.dict). RNA-SeQC provides a particularly detailed coverage metrics report, and it can also compare different samples. The coverage metrics report includes mean coverage, coverage for transcript end regions, bias for 3′ and 5′ ends, and the number, cumulative length, and percentage of gaps. All values are calculated for low, medium, and high expression genes separately. Coverage values are reported also for three levels of GC content. In addition to a coverage uniformity plot, RNA-SeQC plots also coverage over distance (in base pairs) from the 3′ end. The output consists of HTML reports and tab-delimited text files.

Qualimap is a Java program and it uses R and certain Bioconductor packages internally. It takes annotations in GTF/BED format and it needs a separate biotype file as well. Qualimap offers nice plots for saturation and biotype distribution. The saturation plot shows the number of detected features at different sequencing depths, and it also conveniently reports how many new features are detected by increasing the sequencing depth by one million. The biotype distribution plot shows how reads are distributed between protein coding genes, pseudogenes, rRNA, miRNAs, etc., and how big percentage of those features in a genome are covered.

RseQC consists of several Python programs, and it takes genomic annotations in BED format. Note that R needs to be on the path, because it is used internally for plotting results. RseQC has several nice features not found in the other programs: (a) When calculating read distribution between different genomic features, it reports also several bins upstream and downstream of transcripts. (b) Importantly, it calculates the saturation status for splice junctions in addition to genes, and (c) it annotates splice junctions to known, novel and partially novel.

BED files have three obligatory columns and nine optional ones according to the specification [6]. RseQC needs the full 12-column BEDs, because the exon information for each gene is contained in the last three columns (blockCount, blockSizes, and blockStarts). You can obtain BED files for different organisms using the UCSC Table Browser [7]. In the Group menu, select "Genes and gene predictions." The Track menu allows you to choose a gene set, for example, RefSeq genes or Ensembl genes. Set the region to "genome" and the output format to BED. Note that chromosome names in BED files from UCSC contain the prefix "chr," while the alignments produced with Ensembl genomes do not. You can easily remove the chr prefix using the Unix command sed:

```
sed 's/^chr//' hg19_Ensembl_chr.bed>hg19_Ensembl.bed
```

The RseQC example commands below use the TopHat2 paired alignment file `accepted_hits.bam` from Chapter 4.

The tool `read_distribution.py` calculates the distribution of reads to different genomic feature types.

```
python read_distribution.py -r hg19_Ensembl.bed
-i accepted_hits.bam
```

The result table reports total number of reads (excluding nonprimary hits) and tags (separate splice fragments of a read). Total assigned tags indicate how many tags can be assigned unambiguously to the ten different categories listed below.

```
Total Reads                 49743155
Total Tags                  63012643
Total Assigned Tags         57529077
=====================================================
Group           Total_bases     Tag_count     Tags/Kb
CDS_Exons       36821030        34763281      944.11
5'UTR_Exons     34901580        2856644       81.85
```

```
3'UTR_Exons       54908278          9772738        177.98
Introns           1450606807        8468986        5.84
TSS_up_1kb        31234456          94103          3.01
TSS_up_5kb        139129272         161914         1.16
TSS_up_10kb       249300845         217980         0.87
TES_down_1kb      32868738          789703         24.03
TES_down_5kb      142432117         1368378        9.61
TES_down_10kb     251276738         1449448        5.77
=================================================================
```

The tool geneBody_coverage.py produces a coverage plot (Figure 6.1), which allows you to check whether the coverage is uniform along transcripts or if 3′ or 5′ end bias is present. The -o parameter allows you to give a prefix to the result file names.

```
python geneBody_coverage.py -r hg19_Ensembl.bed
-i accepted_hits.bam -o file
```

The precision of gene expression abundance estimates at the current sequencing depth is calculated by the tool RPKM_saturation.py. It resamples subsets of reads, calculates the abundance in RPKM units (described later in this chapter) for each subset, and checks whether they

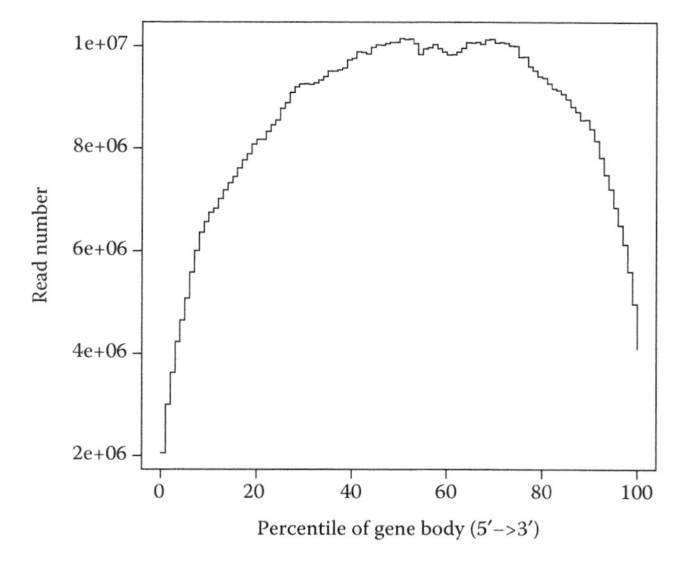

FIGURE 6.1 RseQC plot for coverage uniformity along transcripts. The length of all transcripts is scaled to 100 nucleotides.

are stable or not. This is done separately for four different expression level categories as shown in Figure 6.2.

```
python RPKM_saturation.py -r hg19_Ensembl.bed
-i accepted_hits.bam -o file
```

The tool `junction_annotation.py` divides splice junctions to novel, partially novel (one splice site is novel), and annotated (both splice sites are contained in the reference gene models), and reports the results as a pie chart (Figure 6.3a).

```
python junction_annotation.py -r hg19_Ensembl.bed
-i accepted_hits.bam -o file
```

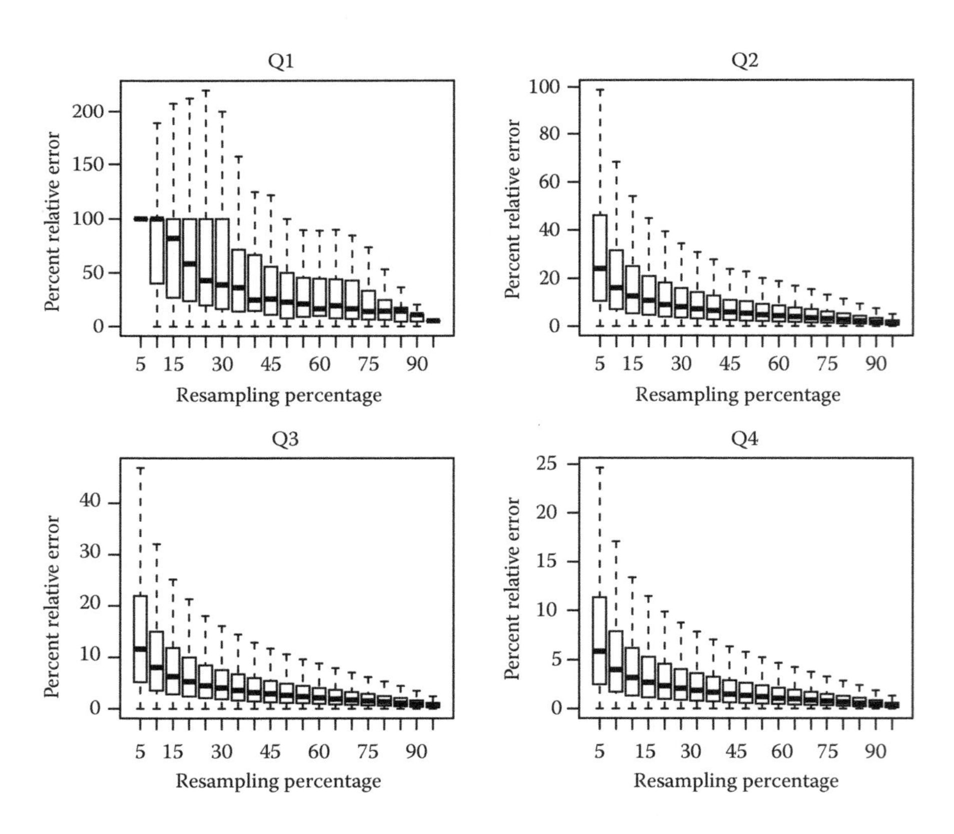

FIGURE 6.2 Sequencing saturation plot by RseQC. Subsets of reads are resampled and RPKMs calculated for each subset and compared to the RPKMs from total reads. This is done separately for four different expression level categories.

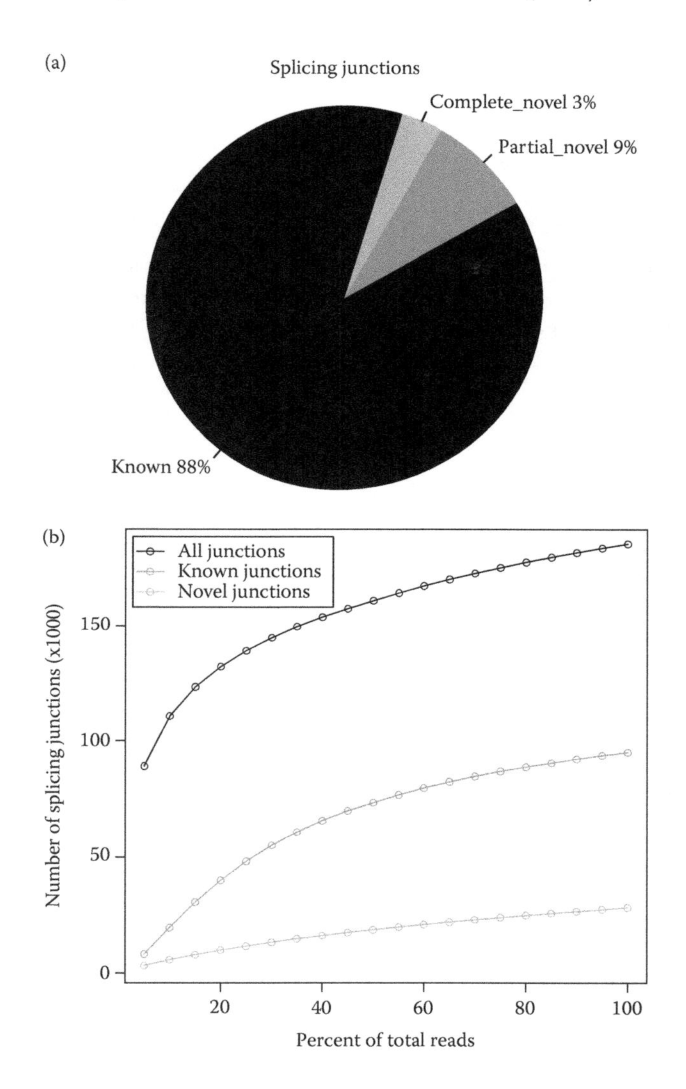

FIGURE 6.3 The RseQC software annotates the detected splice junctions as novel, partially novel, and known (a), and analyzes their saturation status by resampling (b).

Sequencing saturation status for splice junctions can be checked with the tool `junction_saturation.py`. It resamples subsets of reads, detects junctions in each subset, and compares them to the reference annotation. Results are reported for novel and known junctions separately as shown in Figure 6.3b.

```
python junction_saturation.py -r hg19_Ensembl.bed
-i accepted_hits.bam -o file
```

ANNOTATION-BASED QUALITY CONTROL IN CHIPSTER

- Select your alignment file (BAM), a BED file containing annotations, and the tool "Quality control/RNA-seq quality metrics with RseQC." In the parameter panel, make sure that the files have been assigned correctly.

6.3 QUANTITATION OF GENE EXPRESSION

When an annotated reference genome is available, mapped reads can be counted per genomic features based on the location information. Using an annotation file produced by *ab initio* assemblers such as Cufflinks [8] (described in Chapter 5) allows you to quantitate novel genes and transcripts. Alternatively, and especially if there is no reference genome available, reads can be mapped to transcriptome and counted. If there is no reference transcriptome either, you can assemble one using a *de novo* assembler as described in Chapter 5, and then map reads back to this transcriptome for counting.

The number of reads generated per transcript depends on several factors. Some of these are obvious, such as sequencing depth and transcript length (when fragmented during library preparation, longer transcripts produce more fragments and hence more reads). However, some factors affecting the number of reads can be harder to pinpoint, such as transcriptome composition, GC bias, and sequence-specific bias caused by random hexamers. If you want to compare read counts between different genes or different samples, you need to take these factors into account. Many normalization methods are available, and the choice depends on what kind of expression comparisons you want to make. Quantitation software typically outputs abundances in either raw counts or in FPKM (Fragments Per Kilobase per Million mapped reads). Raw counts are needed for differential expression analysis (see Chapter 8 for details), while FPKMs can be used for abundance reporting purposes. FPKM's predecessor RPKM (Reads Per Kilobase per Million mapped reads) was introduced by Mortazavi et al. [9] in order to correct counts for library size and transcript length. It divides counts by transcript length (in kilobases) and by the total number of reads. For example, if a 2 kb transcript has 1000 reads and the total number of reads is 25 million, then RPKM = (1000/2)/25 = 20. FPKM is the equivalent for paired-end experiments where fragments are sequenced from both ends, providing two reads for each fragment. An alternative approach called TPM (Transcripts Per Million) takes into account the distribution of transcript lengths in the sample and should therefore produce abundances which are more consistent between samples [10]. Instead of dividing by the total number of reads, it divides by the sum of "transcript length normalized" reads.

6.3.1 Counting Reads per Genes

The simplest way of estimating expression is to count reads per genes. Several tools are available for this task, such as, HTSeq [11], BEDTools [12], and Qualimap. Also some Bioconductor packages such as Rsubread and GenomicRanges offer counting functionality (code example using GenomicRanges is available in Chapter 7). Also the Cufflinks package provides gene level expression estimates in addition to the transcript ones when assembling transcripts (as described in Chapter 5) and when analyzing differential expression (as described in Chapter 8). All these tools take as input genomic read alignments in SAM/BAM format and genome annotation in GFF/GTF or BED format. They differ in the way how they handle multimapping reads (reads which map to several genomic locations due to homology or sequence repeats): HTSeq ignores these multireads altogether, Qualimap divides the counts equally between the different locations, and Cufflinks has an option to divide each multimapping read probabilistically based on the abundance of the genes it maps to. Counting tools also provide different options for dealing with reads which overlap with more than one gene, or which fall partly in intronic regions. Figure 6.4 illustrates the three counting modes offered by HTSeq, which we use in the examples. All the tools are available for command line use, and Cufflinks, HTSeq, and BEDTools are also available in the Chipster GUI.

6.3.1.1 HTSeq

Htseq-count is part of the HTSeq package of Python scripts for NGS data analysis, but its usage does not require any knowledge of Python. Htseq-count takes aligned reads in the SAM/BAM format and genome

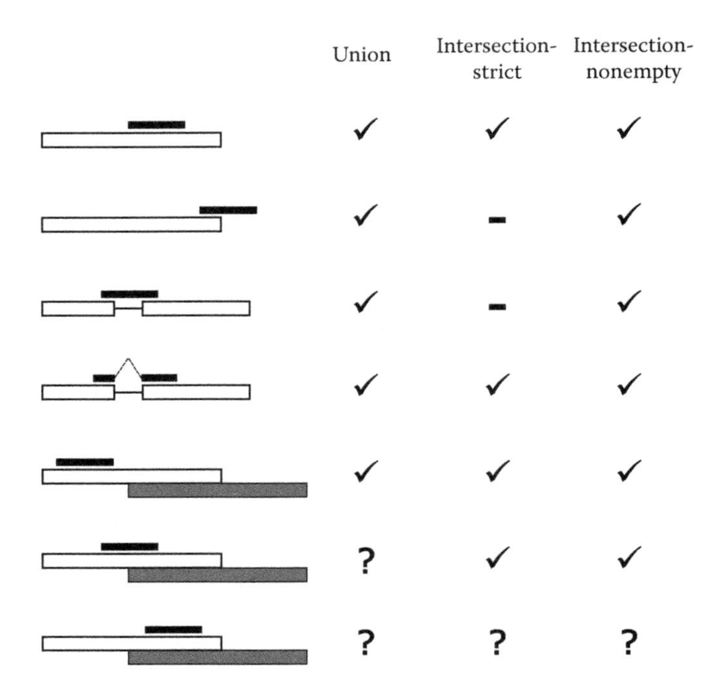

FIGURE 6.4 HTSeq offers three modes to count reads per genomic features. Black bar indicates a read, white box indicates a gene that the read maps to, and the grey box indicates another gene which partially overlaps with the white one. Tick mark means that the read is counted for the white gene, and the question mark means that it is not counted because of ambiguity. The intersection_strict mode does not count the read if it overlaps with intronic or intergenic regions ("no_feature," indicated as dash here). The default setting is the union mode.

annotation as a GFF/GTF file. Note that in order to match the mapping location of reads with genomic features, the alignment file and the annotation file must have the same chromosome names. Htseq-count finds the exons that the reads overlap with and then groups the exon-level counts based on the exons' gene ID in the GTF file. This requires that all the exons of a particular gene have the same gene ID. While Ensembl GTF files follow this rule, the GTF files available in the UCSC Table browser have a transcript ID repeated as a gene ID. This is problematic to htseq-count, because it cannot guess which transcripts belong to the same gene, and hence it will count the reads separately. Ensembl GTFs are available at http://www.ensembl.org/info/data/ftp/index.html

by selecting the organism and the option "GTF." In the examples below, we use the TopHat2 paired alignment file from Chapter 4, so we want to download the human GTF:

```
wget ftp://ftp.ensembl.org/pub/release74/gtf
/homo_sapiens/Homo_sapiens.GRCh37.74.gtf.gz
```

Unzip the file

```
gunzip Homo_sapiens.GRCh37.74.gtf.gz
```

By default htseq-count expects paired-end data to be sorted by read names so that the paired reads follow each other in the file. The alignments can also be sorted by genomic position (use the option `-order=pos`), but this has higher memory requirements. The following command sorts BAM by read names and produces a file hits_namesorted.bam

```
samtools sort -n accepted_hits.bam hits_namesorted
```

The htseq-count command looks like this (make sure that the file htseq-qa is on the path):

```
htseq-count -f bam --stranded=no hits_namesorted.bam
Homo_sapiens.GRCh37.74.gtf> counts.txt
```

Here `-f bam` indicates that the input format is BAM. The default behavior is to count reads which match the exon locations in the GTF file (`--type=exon`) and combine counts for exons that belong to the same gene (`--idattr=gene_id`). Htseq-count assumes that data were produced with a strand-specific protocol and it counts reads only if they map to the same strand as the gene. As the example data are not stranded, we have to add `--stranded=no` so that a read is counted also when it maps to the opposite strand. The default counting mode is union, but you can change that with the `--mode` option, and you can also set a minimum mapping quality for a read to be counted (e.g., `–a 30`), the default is 10.

The output counts.txt is a table of counts for each gene. In the end of the file, there are five rows listing the number of reads which were not counted for any gene because

a. based on the NH tag in the BAM file, they aligned to more than one place in the reference genome (alignment_not_unique);

b. they did not align at all (not_aligned);

c. their alignment quality was lower than the user-specified threshold (too_low_aQual);

d. their alignment overlapped with more than one gene (ambiguous);

e. their alignment did not overlap any gene (no_feature).

```
...
ENSG00000273490    0
ENSG00000273491    0
ENSG00000273492    0
ENSG00000273493    0
_no_feature        6125428
_ambiguous         1808462
_too_low_aQual     0
_not_aligned       0
_alignment_not_unique 2947054
```

You can combine count files from different samples to a table using the Unix command join:

```
join counts1.txt counts2.txt > count_table.txt
```

Finally, you might like to remove the last five rows prior to statistical testing for differential expression. The following Unix command head keeps all but the last five lines (`-n -5`):

```
head -n -5 count_table.txt > genecounts.txt
```

6.3.2 Counting Reads per Transcripts

Counting reads at the transcript level is complicated by the fact that transcript isoforms typically have overlapping parts. In order to assign ambiguously mapping reads to different isoforms, an expectation maximization (EM) approach is used. This approach alternates between two steps: an expectation step where reads are assigned to transcripts with a probability according to those transcripts' abundances (which are

COUNTING READS PER GENES IN CHIPSTER

- Select your alignment file (BAM) and the tool "RNA-seq/Count aligned reads per genes with HTSeq." In the parameters, select the organism and indicate whether your data were produced with a strand-specific protocol. You can also choose to include genes' chromosomal coordinates in the count file (this helps visualizing differential expression analysis results in a genome browser later on). Click "Run."
- Note that if the organism of your sample is not available in Chipster, you can use the tool "RNA-seq/Count aligned reads per genes with HTSeq using own GTF." Import the GTF file to Chipster and select it together with the BAM file as input. In the parameter window, make sure that the files have been assigned correctly.
- Select the count files for all the samples and combine them to a count table using the tool "Utilities/Define NGS experiment." In the parameters, indicate the column-containing counts and whether your data contain chromosomal coordinates or not.

initially assumed to be equal), and a maximization step where the abundances are updated based on the assignment probability. Programs that estimate transcript abundances in multi-isoform genes this way include Cufflinks and eXpress [13]. Cufflinks uses a batch EM approach, while eXpress uses an online EM algorithm and is therefore faster and more memory efficient.

While Cufflinks uses genomic alignments as input, eXpress uses alignments to transcriptome and is thus suitable also for species which do not have a reference genome yet. If reference transcriptome is not available either, you can create it with a *de novo* assembler such as Trinity or Oases as described in Chapter 5. Abundance estimates produced by eXpress can be efficiently updated with the ReXpress tool [14] when transcript annotations change. Avoiding time-consuming reanalysis of the whole data set is particularly important for newly sequenced organisms, whose transcript annotations change often.

Both Cufflinks and eXpress can resolve multimappings of reads across gene families, learn fragment length distribution from data, and correct for sequence-specific bias near the ends of fragments, which arises due to primers used in library preparation. In addition, eXpress also includes a model for sequencing errors including indels, and it can estimate allele-specific expression. In addition to command line, Cufflinks and eXpress can be used in the Chipster GUI.

6.3.2.1 Cufflinks

Cufflinks takes genomic alignments in BAM format and annotations in GTF file. The GTF file is optional, because Cufflinks can combine isoform abundance estimation and assembly. Using fragment bias correction is recommended. When it is enabled, Cufflinks learns from the data what sequences were selected for, and re-estimates the abundances with a new likelihood function that takes the sequence-specific bias into account (Cufflinks uses the original abundance information in order to make a difference between sequences which are common due to high expression rather than bias).

The example Cufflinks command takes as input the paired-end genomic alignment produced by TopHat2. It estimates the expression of known transcripts and does not assemble novel ones (`-G`). It corrects for fragment bias (`-b GRCh37.74.fa`) and weighs reads which map to multiple locations (`-u`). Eight processors are used to speed up the process (`-p 8`). Note that SAMtools needs to be on the path, because Cufflinks uses it internally.

```
cufflinks -G Homo_sapiens.GRCh37.74.gtf -b GRCh37.74.
fa -u -p 8 accepted_hits.bam -o outputFolder
```

The output consists of transcript and gene-level FPKM-tracking files, which contain FPKM values and their confidence intervals. FPKM-tracking files are also produced when a set of samples is tested for differential expression using Cuffdiff, as described in Chapter 8.

6.3.2.2 eXpress

eXpress takes as input transcript sequences in multi-FASTA format and read alignments which were made using this set of transcripts. The alignments can be in a BAM file or they can be streamed directly from the aligner such as Bowtie2 to eXpress (spliced aligner is not needed because reads are mapped to transcriptome instead of genome). It is important to allow as many multimappings as possible. You can also allow many mismatches, because eXpress builds an error model to probabilistically assign the reads. BAM/SAM files containing paired-end data need to be sorted by read names as described in the HTSeq section above.

In the following example, we download transcript sequences from the RefSeq database [15], create a Bowtie2 index for this set, align reads with Bowtie2, and calculate transcript abundances using eXpress.

Download transcripts from RefSeq:

```
wget ftp://ftp.ncbi.nlm.nih.gov/refseq/H_sapiens/
mRNA_Prot/human.rna.fna.gz
```

Uncompress the file:

```
gunzip human.rna.fna.gz
```

Rename it so that you remember the RefSeq version used:

```
mv human.rna.fna refseq63.fasta
```

Create Bowtie2 index for the transcripts as described in Chapter 4. The off-rate parameter controls how many rows in the reference index are marked. The default value is 5, which means that every 32nd row ($=2^5$) is marked. We change it to 1 so that every second row is marked in order to make reference position lookups faster during the alignment. This is necessary because we want to allow many multimappings during alignment, which makes Bowtie2 really slow.

```
bowtie2-build -offrate=1 -f refseq63.fasta refseq63
```

The following command aligns reads to transcripts using Bowtie2 parameter settings recommended by the eXpress authors (http://bio.math.berkeley.edu/ReXpress/rexpress_manual.html). We use the `-k` option to tell Bowtie2 to report 1000 alignments per read, instead of just one. Ideally, we would like to have all alignments (`-a`), but this would be even slower because Bowtie2 was not designed for this kind of use. The SAM output from Bowtie2 is piped to SAMtools for conversion to BAM in order to save space (the SAM file produced by Bowtie2 is automatically name-sorted, so we omit the sorting step here).

```
bowtie2 -q -k 1000 -p 8 --phred64 --no-discordant
--no-mixed --rdg 6,5 --rfg 6,5 --score-min L,-.6,-.4
-x refseq63 -1 reads1.fq.gz -2 reads2.fq.gz | samtools
view -Sb - >transcriptome_aligned.bam
```

The search is restricted to concordant, paired alignments (`--no-discordant --no-mixed`) and it is made stricter by increasing the read and reference gap penalties (`--rdg 6,5 --rfg 6,5`) and the minimum accepted alignment score (--score-min L,-.6,-.4) from their default values. The overall alignment rate was 69.17% as indicated by the screen summary:

```
34232081 reads; of these:
 34232081 (100.00%) were paired; of these:
  10553741 (30.83%) aligned concordantly 0 times
  4166418 (12.17%) aligned concordantly exactly 1 time
  19511922 (57.00%) aligned concordantly >1 times
69.17% overall alignment rate
```

Calculate transcript abundances with eXpress using bias correction and error correction:

```
express refseq63.fasta transcriptome_aligned.bam -o
outputFolder
```

Alternatively, you can pipe the Bowtie2 output directly into eXpress in order to avoid the large intermediate BAM file:

```
bowtie2 -k 1000 -p 8 --phred64 --no-discordant --no-mixed
--rdg 6,5 --rfg 6,5 --score-min L,-.6,-.4 -x refseq63 -1
reads_1.fq.gz -2 reads_2.fq.gz | express refseq63.fasta
-o outputFolder
```

The result file results.xprs contains the abundance estimates. Transcripts are sorted by a bundle (bundle_id), which is defined as a group of transcripts that share multimapping reads. The file has several columns, and the most important ones are estimated counts (est_counts), effective counts (eff_counts), FPKM, and TPM. Effective counts are adjusted for fragment and length biases, and the authors of eXpress recommend using them rounded for count-based differential expression analysis tools like edgeR. The following awk command extracts the transcript identifier and the effective counts column:

```
awk '{print$2"\t"$8}'results.xprs>eff_counts.txt
```

The beginning of the result file looks like this:

```
target_id                           eff_counts
gi|530366287|ref|XM_005273173.1|    0.000000
gi|223555918|ref|NM_152415.2|       463.539280
```

```
gi|530387564|ref|XM_005273400.1|     0.481096
gi|530387566|ref|XM_005273401.1|     25.786556
gi|223555920|ref|NM_001145152.1|     9.204109
gi|225543473|ref|NM_004686.4|        28.171057
```

You can keep only the RefSeq identifiers and trim the decimals using the following awk command, where the option -F specifies the field separator (here |). The first line is copied as is (NR==1{print;next}). For the next lines, only the fourth and fifth fields are kept, and the numbers in the fifth field are rounded.

```
awk -F'|''NR==1 {print;next}
{print$4"\t"int($5+0.5)}'eff_counts.txt>eff_counts_
rounded.txt
```

The beginning of the result file looks like this.

```
target_id          eff_counts
XM_005273173.1     0
NM_152415.2        464
XM_005273400.1     0
XM_005273401.1     26
NM_001145152.1     9
NM_004686.4        28
```

You can check how many transcripts have rounded effective counts using the following command. It uses awk to collect lines where the value in the second column is different from zero, and the result is piped to the Unix command wc -l which counts how many lines there are.

```
awk '$2!=0{print}'eff_counts_rounded.txt | wc -l
```

According to this, 52,259 transcripts (out of the 91,950 measured) had effective counts. You need to sort the data by the identifier column in order to combine count files from different samples to a count table later. The following commands extract the title row and append the sorted data to it:

```
head -n 1 eff_counts_rounded.txt > eff_counts_rounded_
sorted.txt
```

```
tail -n +2 eff_counts_rounded.txt | sort -k 1,1 >>
eff_counts_rounded_sorted.txt
```

COUNTING READS PER TRANSCRIPTS IN CHIPSTER

You can use the tool "RNA-seq/Assemble reads to transcripts with Cufflinks" as described in Chapter 5. You can also use eXpress:

- Select the FASTQ file(s), the multi-FASTA file containing the transcript sequences, and the tool "RNA-seq / Count reads per transcripts using eXpress." In the parameter window, make sure that the files have been assigned correctly.
- Select the count files for all the samples and combine them to a count table using the tool "Utilities/Define NGS experiment." In the parameters, select the column containing counts and indicate that your data do not contain chromosomal coordinates.

6.3.3 Counting Reads per Exons

Differential expression can be studied at exon level using the Bioconductor package DEXSeq [16] as described in Chapter 9 and for that we need to count reads per exons. Transcript isoforms tend to have some exons in common, so an exon can appear several times in a GTF file. Exons can also overlap with each other if their start/end coordinates differ. For counting purposes, we need to construct a set of nonoverlapping exonic regions. The DEXSeq package contains a Python script dexseq_prepare_annotation.py for this task. It "flattens" a GTF file to a list of exon counting bins, which correspond to one exon or a part of an exon (in the case of overlap). As discussed in the context of HTSeq above, it is important to use a GTF file where all the exons of a gene have the same gene ID. Ensembl GTF files are recommended, because they follow this rule. The following examples use the TopHat2 paired alignment file from Chapter 4.

Download an Ensembl GTF file for human as shown in the HTSeq section above and "flatten" it with the following command:

```
python dexseq_prepare_annotation.py Homo_sapiens.
GRCh37.74.gtf GRCh37.74_DEX.gtf
```

The Python script dexseq_count.py contained in the DEXSeq package is used to count reads per nonoverlapping exonic parts. It takes as input the "flattened" GTF file and aligned reads in SAM format. BAM can be used as well, but you have to install the Python package Pysam [17] for this to work. The following command indicates that our data are paired-end (`-p yes`) and sorted by read names (`-r name`). The script would also accept data sorted by chromosomal coordinates (-r pos). We need to indicate that our data are not stranded (-s no), because the script assumes that the data were produced with a strand-specific protocol. It is also possible to set a mapping quality threshold for reads to be counted (e.g., -a 30), the default is 10.

```
python dexseq_count.py -p yes -s no -r name GRCh37.74_
DEX.gtf hits_namesorted.sam exon_counts.txt
```

The count file lists the number of reads for each exon counting bin. Bin identifiers consist of gene identifiers followed by an exon bin number. Some bin identifiers have two gene identifiers separated by a plus sign as shown below. This means that the two genes are on the same strand and their exons overlap.

```
ENSG00000001036:001       210
ENSG00000001036:002       12
ENSG00000001036:003       6
ENSG00000001036:004       135
ENSG00000001036:005       82
ENSG00000001036:006       205
ENSG00000001036:007       138
ENSG00000001036:008       2
ENSG00000001036:009       21
ENSG00000001036:010       76
ENSG00000001036:011       25
ENSG00000001084+ENSG00000231683:001    57
ENSG00000001084+ENSG00000231683:002    57
ENSG00000001084+ENSG00000231683:003    50
ENSG00000001084+ENSG00000231683:004    34
```

The last four rows of the file list the number of reads that were not counted because

a. they were not aligned at all (__notaligned)

b. the alignment quality was lower than the user-specified threshold (__lowaqual)

c. the alignment overlapped with more than one exon counting bin (__ambiguous)

d. the alignment didn't overlap with any exon counting bin (__empty)

You can remove these lines as described in the HTSeq section above (head `-n -4`). Finally, you can combine counts from different samples to a count table using the Unix command `join` as shown in the HTSeq section, although it is not necessary for DEXseq.

COUNTING READS PER EXONS IN CHIPSTER

- Select your alignment file (BAM) and the tool "RNA-seq/Count aligned reads per exons for DEXSeq." In the parameters, select the organism, and indicate whether your data were produced with a paired-end or strand-specific protocol. Click "Run."
- Select the count files for all the samples and combine them to a count table using the tool "Utilities/Define NGS experiment." In the parameters, select the column containing counts and indicate that your data do not contain chromosomal coordinates.

6.4 SUMMARY

Matching the genomic locations of aligned reads with reference annotation allows you to investigate important quality aspects, such as saturation of sequencing depth, coverage uniformity along transcripts, and read distribution between different genomic feature types. Several tools are available for annotation-based quality control and they all have their particular advantages.

When reads have been mapped to a reference, we can also quantitate gene expression by counting reads per genes, transcripts, and exons. Quantitation and differential expression analysis are inherently interlinked, and the best practices are still being debated. Reads can be counted per genes with tools like HTSeq, but gene-level counts are not optimal for differential expression analysis of those genes which undergo isoform switching (because longer transcripts give more counts). The major challenge in quantitating expression at transcript level is how to assign ambiguously mapping reads to different isoforms. Cufflinks and eXpress apply an EM approach for this task. Cufflinks needs a reference genome, while eXpress uses transcriptome alignments, and can hence be used also for organisms which do not have a reference genome available. Quantitating expression at isoform level is challenging also because transcript coverage is typically not uniform due to mappability issues and biases introduced in library preparation and sequencing. For abundance reporting purposes, counts can be normalized for library size and transcript length using units like FPKM and TPM. Differential expression analysis typically uses raw counts and applies an internal normalization procedure in order to account for differences in transcriptome composition.

REFERENCES

1. Wang L., Wang S., and Li W. RSeQC: Quality control of RNA-seq experiments. *Bioinformatics* **28**(16):2184–2185, 2012.
2. DeLuca D.S., Levin J.Z., Sivachenko A. et al. RNA-SeQC: RNA-seq metrics for quality control and process optimization. *Bioinformatics* **28**(11):1530–1532, 2012.
3. Garcia-Alcalde F., Okonechnikov K., Carbonell J. et al. Qualimap: Evaluating next-generation sequencing alignment data. *Bioinformatics* **28**(20):2678–2679, 2012.
4. *Picard*. Available from: http://picard.sourceforge.net/.
5. *GFF/GTF file format description*. Available from: http://genome.ucsc.edu/FAQ/FAQformat.html#format3.
6. *BED file format description*. Available from: http://genome.ucsc.edu/FAQ/FAQformat.html#format1.
7. *UCSC Table Browser*. Available from: http://genome.ucsc.edu/cgi-bin/hgTables.
8. Trapnell C., Williams B.A., Pertea G. et al. Transcript assembly and quantification by RNA-seq reveals unannotated transcripts and isoform switching during cell differentiation. *Nat Biotechnol* **28**(5):511–515, 2010.
9. Mortazavi A., Williams B.A., McCue K., Schaeffer L., and Wold B. Mapping and quantifying mammalian transcriptomes by RNA-seq. *Nat Methods* **5**(7):621–628, 2008.
10. Wagner G.P., Kin K., and Lynch V.J. Measurement of mRNA abundance using RNA-seq data: RPKM measure is inconsistent among samples. *Theory Biosci* **131**(4):281–285, 2012.
11. Anders S., Pyl P.T., and Huber, W. HTSeq – A Python framework to work with high-throughput sequencing data. bioRxiv doi: 10.1101/002824, 2014.
12. Quinlan A.R. and Hall I.M. BEDTools: A flexible suite of utilities for comparing genomic features. *Bioinformatics* **26**(6):841–842, 2010.
13. Roberts A. and Pachter L. Streaming fragment assignment for real-time analysis of sequencing experiments. *Nat Methods* **10**(1):71–73, 2013.
14. Roberts A., Schaeffer L., and Pachter L. Updating RNA-seq analyses after re-annotation. *Bioinformatics* **29**(13):1631–1637, 2013.
15. Pruitt K.D., Tatusova T., Brown G.R., and Maglott D.R. NCBI Reference Sequences (RefSeq): Current status, new features and genome annotation policy. *Nucleic Acids Res* **40**(Database issue):D130–D135, 2012.
16. Anders S., Reyes A., and Huber W. Detecting differential usage of exons from RNA-seq data. *Genome Res* **22**(10):2008–2017, 2012.
17. *Pysam*. Available from: https://code.google.com/p/pysam/.

CHAPTER 7

RNA-seq Analysis Framework in R and Bioconductor

7.1 INTRODUCTION

R (R Core Team [1]; http://www.r-project.org) is an open source software for statistical programming, an analysis environment, and a community formed by the users and the developers of the software. R software consists of a core and thousands of optional add-on packages that extend the functionality of the core. R core is developed by the R core development team, but most of the add-on packages are contributed by third-party developers, such as academic researchers from various universities around the world. Bioconductor ([2]; http://www.bioconductor.org) is a large software development project that provides tools for genomic and high-throughput data analysis. Software developed in the Bioconductor project is released as R add-on packages.

R is distinctively a programming language for statistics, data mining, and also bioinformatics. It differs from many other programming languages in its heavy emphasis of the statistical functionality. There are some other languages, such as Python, that offer comprehensive computational and statistical functions, but R has a special role in the community, because it sees many of the bleeding edge developments before other languages. In the field of statistics, R can be somewhat compared with, for example, SAS and Stata, both containing a programming or scripting language with which the analyses are performed. For the basic statistical or bioinformatics work, the knowledge of all the programming nuances of the R language is not needed, and one can perform the analyses successfully (to some extent) just by getting to know the most commonly used functions. However, delving deeper into the language will help with the more difficult analyses or with the various data manipulation steps that can sometimes get rather complex.

This chapter offers an overview of the R and Bioconductor functionality for high-throughput sequencing analyses. If you need to get acquainted with R functionality, consider studying the manuals at http://cran.r-project.org/manuals.html. The same manuals also ship with the R installation. On top of these basic guides, there are a number of introductory books; among others the excellent R in action by Kabacoff from Manning, Rizzo's Statistical computing with R from the CRC Press, and the more timid, but very broad R in a nutshell by Adler from O'Reilly.

7.1.1 Installing R and Add-on Packages

R can be installed from the comprehensive R archive network (CRAN; http://cran.at.r-project.org/) or from any of its mirrors around the world. A link to CRAN mirrors is found on the main page of the R project under the heading "Download, Packages." From one of the CRAN mirror servers, you need to download and install the R base. A direct link to the base R for Windows download page on the main CRAN mirror at Austria is http://cran.at.r-project.org/bin/windows/base/. Download the installer, run it, and follow the instructions given by the installer. If your institute does not allow you to install the software on your workstation on your own, please consult the local IT support and direct them to the pages mentioned above.

Once you have installed the base R, you can typically install add-on packages directly from R. Getting to know the packages that are needed is the part that requires some research. A browsable list of CRAN packages is available at http://cran.at.r-project.org/web/packages/available_packages_by_name.html. The list contains short descriptions of the functionalities of each of the packages. In addition, there is a rather comprehensive task-based grouping of the packages, known as the task views, at http://cran.at.r-project.org/web/views/. A description of the Bioconductor packages is available at http://www.bioconductor.org/packages/release/BiocViews.html.

Once you have identified the packages you will need, they can be installed as follows:

1. For the CRAN packages, you can go to the Packages menu in the R program and select the "Install Package(s) ..." functionality. You need to select the CRAN mirror you want to install from, and the package(s) you would like to install. After that R will automatically download the packages and install them.
2. The Bioconductor packages can be installed similarly to the CRAN packages, but the suggested method is to first load the helper function biocLite() from the Bioconductor site. Just type `source("http://www.bioconductor.org/biocLite.R")` on the R command line and press the Enter key to execute the command. Once the helper function has been loaded, you can install Bioconductor for the first time just by giving the command `biocLite()`. Individual packages can be installed by giving them as argument to the helper function. For example, the Gviz package for genomic visualization can be installed with the command `biocLite("Gviz")`.

Sometimes packages cannot be installed directly, because the network firewall blocks the connections to the CRAN mirrors. Often the situation can be rectified on Windows machines by giving the command `setInternet2()` which allows R to make use of the Internet Explorer functions, such as specification of the proxies.

7.1.2 Using R

R is a command line tool. Windows and Mac OS X offer a simple GUI to R, but on Linux (and UNIX) machines the command line is the only user interface. There are some graphical user interfaces to R, such as R Commander, and many more development environments and code editors, such as R Studio and TinnR. There are even graphical programming environments for R, such as that offered by the Alteryx. However, the access to all functionality is most comprehensively available if R is used from the command line.

Each new line in the R editor starts with the prompt, which is a simple character ">." The commands and functions are written to the prompt and then executed by pressing the Enter key from the keyboard. The key to using R successfully is of course to get to know what to type on the prompt. The aim of the following chapters in this book is to give an idea of how some type of analyses can be performed in R. However, this is not a basic book on R and at least some previous knowledge of R is required to successfully apply the ideas presented in the book.

When you encounter code lines in this book, consider running them one line at a time and observing what happens when you have executed the line. In addition, it is a good habit to consult the help for the new functions you do not know beforehand. The help page for the function can be invoked with the function `?` of `help()`. For example, the help page for the function `lm()` can be invoked by giving the command `?lm`.

7.2 OVERVIEW OF THE BIOCONDUCTOR PACKAGES

The add-on packages produced by the Bioconductor project can be broadly divided into software, annotation, and experiment packages. Software packages contain the analysis functionality, annotation packages of various types of annotations, and the experiment packages contain the data sets that are often used as examples of the package functionalities. Let us take a slightly more detailed look at these package categories.

7.2.1 Software Packages

In general, the Bioconductor software packages contain functionality for importing, manipulating (preprocessing and quality controlling), analyzing, plotting, and reporting the results from high-throughput experiments.

For RNA-seq experiments, the most important packages are (1) Short Read and Rsamtools for reading and writing sequence files, (2) IRanges, GenomicRanges, and Biostrings for data manipulation, (3) edgeR, DESeq, and DEXSeq for statistical analyses, (4) rtracklayer, BSgenome, and biomaRt for annotating the results.

7.2.2 Annotation Packages

The Bioconductor project produces basic annotation packages for many organisms. These annotation packages can be divided into genome sequence (BSgenome. packages), genome-wide annotation (org. packages), transcript (TxDB. packages), homology (hom. packages), microRNA target (RmiR. and targetscan. packages), functional annotation (DO, GO, KEGG, reactome) packages, variants (SNPlocs. packages), and predictions of variant functions (SIFT. and PolyPhen. packages). These packages typically offer the annotations from the US sources, such as Genbank and UCSC, and the accession numbers, for example, for genes are taken from the Entrez Gene. However, these packages offer the functionality to translate the Entrez Gene IDs, for example, to Ensembl IDs, usually via the organism-specific org-package.

In addition to the readymade annotation packages, annotations can be queried directly from the online sources. The Bioconductor package biomaRt allows the user to access the whole BioMart genome data warehouse. Similarly, rtracklayer package allows one to query UCSC genome browser's annotation tracks. In addition, packages arrayexpress and GEOquery connect R with the ArrayExpress and GEO databases.

7.2.3 Experiment Packages

Experiment packages contain ready-packaged, freely available data sets. In this book, the parathyroid data set from the similarly named package is used for demonstrating the statistical analyses using DESeq and DEXSeq software packages.

7.3 DESCRIPTIVE FEATURES OF THE BIOCONDUCTOR PACKAGES

Bioconductor packages employ the object-oriented programming (OOP) paradigm extensively. In R, OOP is rationalized through methods that work on S3 and S4 object classes. S3 simulates only certain aspects of OOP, but S4 is a formal OOP system, the so-called fourth version of the S language of which R is the open source implementation. Each class extends one or more classes, and in comparison to Java classes, S4 classes do not own the methods. Typically, there is a generic function that selects a specialized function for a certain set of functions. Specialized functions are also called methods. The OOP system implemented in R is described by Chambers [3,4].

7.3.1 OOP Features in R

Where there is a function in base R, there is often a method in OOB. Similarly, when a table (matrix or data frame) or a list is used in base R, an S3/S4 object is used in OOB. S3 and S4 classes of objects contain slots that store different types of data. A single column of a data frame could be accessed using the $ operator, but for the S3/S4 classes of objects, individual slots are accessed using the @ operator. It is better to use an accessor function rather than the @-operator to extract a slot from an S4 object, because using the accessor function is independent of class representation. If the name of the slot changes, the @-operator would cease to work, but the accessor function, if appropriately updated by the package developer, will continue to work correctly.

To make this a bit more concrete, let us represent a single gene as a sequence range object. We can use the package GenomicRanges for that. A new sequence range object is created using the function `GRanges()`. The following code creates a representation of the XRCC1 gene that is located on the forward strand of the chromosome 19 between the locations 44047464 and 44047499:

```
library(GenomicRanges)
read<-GRanges(seqnames=c("19"),
                 ranges=IRanges(start=c(44047464),
                 end=c(44047499)), strand=c("+"),
                 seqlenghts=c("19"=591289983))
names(read)<-c("XRCC1")
```

If the contents of the object read is checked with the function `str()`, the output should appear as follows. Can you locate the information you inputted for the gene XRCC1 from the output?

```
str(read)

Formal class:'GRanges ' [package "GenomicRanges"] with 6
slots
  .. @ seqnames: Formal class 'Rle'[package "IRanges"]
  with 4 slots
  ......@ values          : Factor w/1 level "19":1
  ......@ lengths         : int 1
  ......@ elementMetadata: NULL
  ......@ metadata        : list()
  ..@ ranges: Formal class 'IRanges' [package
  "IRanges"]with 6 slots
  ......@ start           : int 44047464
  ......@ width           : int 36
  ......@ NAMES           : chr "XRCC1"
  ......@ elementType     : chr "integer"
  ......@ elementMetadata: NULL
  ......@ metadata        : list()
  ..@ strand: Formal class 'Rle' [package "IRanges"]
  with 4 slots
  ......@ values          : Factor w/3 levels "+",
  "-","*":1
```

```
......@ lengths         : int 1
......@ elementMetadata: NULL
......@ metadata        : list()
..@ elementMetadata: Formal class 'DataFrame'
[package "IRanges"] with 6 slots
......@ rownames        : NULL
......@ nrows           : int 1
......@ listData        : List of 1
........$ seqlenghts:Named num 59128983
..........- attr(*,"names")=chr "19"
......@ elementType     : chr "ANY"
......@ elementMetadata: NULL
......@ metadata        : list()
.. @ seqinfo: Formal class 'Seqinfo' [package
"GenomicRanges"] with 4 slots
......@ seqnames        : chr "19"
......@ seqlengths      : int NA
......@ is_circular     : logi NA
......@ genome          : chr NA
.. @ metadata           : list()
```

All of the object's slots are preceded with the @-sign, and they can be accessed using the same operator. For example, NAMES slot can be extracted with the command:

```
read@ranges@NAMES
[1]"XRCC1"
```

However, there are also accessor functions for the most important slots of the objects. For example, the sequence name can also be accessed using the function `names()`:

```
names(read)
[1]"XRCC1"
```

Sometimes the S3/S4 objects can be coerced to other object types directly, but this is not typically always possible. For example, for writing a GRanges object as a table on the disc, it can be first converted into a data frame:

```
read.df <-as.data.frame(read)
read.df
     seqnames    start      end width strand seqlenghts
XRCC1       19 44047464 44047499    36      +   59128983
```

Writing a data frame to a disc could then be accomplished via the usual route of `write.table()`. The individual columns of the data frame can be accessed using the dollar-operator. For example, the width of the single read stored in the read.df data frame can be printed to the screen with:

```
read.df$width
[1]36
```

Similarly, the names of the sequences stored on the lines can be accessed with the function

```
rownames():
rownames(read.df)
[1]"XRCC1"
```

7.4 REPRESENTING GENES AND TRANSCRIPTS IN R

Genes and transcripts are typically represented as sequence ranges (e.g., Granges objects) in R, since they have some nice computational properties; among others, they take little space and can be handled with fast algorithms. The main workhorse packages are IRanges and GenomicRanges which build on the IRanges package.

Rsamtools is a package that makes the functionality of samtools available in R. For example, it can be used to read the BAM files to R. The BAM file is converted into an object of Granges class, which contains all the individual reads from the BAM file. To read a single BAM file to R, function `readGappedAlignments()` can be used, for example:

```
library(Rsamtools)
h1b<-readBamGappedAlignments("hESC1_chr18.bam")
```

Object h1b contains the aligned sequence from a BAM file. The key information for each read is the chromosome, strand, and the base pair location of the mapped read in the genome. A short stretch from the beginning and the end of the R object h1b can be printed on the screen just by typing the name of the object to the prompt:

```
h1b

GappedAlignments with 836162 alignments and 0 metadata
columns:
```

```
           seq names strand cigar <char-qwidth    start      end        width
           <Rle>     <Rle>  acter>        <integer> <integer> <integer> <integer>
      [1]     chr18     +        75M        75    28842     28916        75
      [2]     chr18     +        75M        75    35847     35921        75
      [3]     chr18     -        75M        75    46570     46644        75
      [4]     chr18     -        75M        75    46570     46644        75
      [5]     chr18     +        75M        75    47246     47320        75
      ...       ...   ...        ...       ...      ...       ...       ...
 [836158]     chr18     +        75M        75 78005301  78005375        75
 [836159]    chr18-     -        75M        75 78005301  78005375        75
 [836160]     chr18     +        75M        75 78005307  78005381        75
 [836161]    chr18-     -        75M        75 78005309  78005383        75
 [836162]    chr18-     -        75M        75 78005366  78005440        75


             ngap <integer>
      [1]                 0
      [2]                 0
      [3]                 0
      [4]                 0
      [5]                 0
      ...               ...
 [836158]                 0
 [836159]                 0
 [836160]                 0
 [836161]                 0
 [836162]                 0
---
seqlengths:
     chr1      chr2      chr3      chr4 ...    chr22      chrX  chrM
249250621 243199373 198022430 191154276 ... 51304566 155270560 16571
```

The Range-object such as those of the GenomicRanges class can be subsetted, if needed, using the usual way subsetting the object with the square brackets. For example, the first ten genes can be displayed on the screen with:

```
h1b[1:10,]
```

Similarly, the genes only on the forward strand can be extracted:

```
h1b[strand(h1b)=="+",]
```

Once the aligned sequences from the BAM file are loaded into a range-based object, we can easily count how many reads map to each gene. Transcripts for certain model organisms are readily available as R packages. For example, the transcript from the hg19 assembly of the human genome taken from the UCSC's Genome server is available in the Bioconductor package `TxDb.Hsapiens.UCSC.hg19.knownGene`. Genes can be extracted as ranges from the package into a new object txdb:

```
library(TxDb.Hsapiens.UCSC.hg19.knownGene)
txdb<-transcriptsBy(TxDb.Hsapiens.UCSC.hg19.knownGene,
                    "gene")
```

Instead of genes, we could also extract exons, coding sequences, or transcripts with the same function. Once the desired genomic features are extracted, we can count the number of reads (or their ranges) that overlap the genomic features (or their ranges). The function `countOverlaps()` takes care of the details of the counting:

```
hits <- countOverlaps(h1b, txdb)
ol<-countOverlaps(txdb, h1b[hits==1])
```

The first run of the `countOverlaps()` marks the individual reads with running numbers. The reads with the flag one map to a single gene only and these are mapped to the genes they map to during the second run of the function `countOverlaps()`. The result object ol is a named numeric vector, where the names are Entrez Gene identifiers, and the numbers are counts of reads that overlap the named gene.

Using the functions just introduced, we can cook up a function that reads a number of BAM files, counts the number of reads that map to each gene, and gives out a count table. Naturally, this only works correctly, if the mapping is done against the same assembly version of the genome that is available from the Bioconductor project. The code for the function is below:

```
generateCountTable <- function(
   files,
   transcripts="TxDb.Hsapiens.UCSC.hg19.knownGene",
   overlapto="gene") {
      require(transcripts, character.only=TRUE)
      require(GenomicRanges)
      require(Rsamtools)
      txdb<-transcriptsBy(get(transcripts,
                             envir=.GlobalEnv),
                             overlapto)
      l<- vector("list", length(files))
```

```
        for(i in 1:length(files)) {
            alns <- readGappedAlignments(files[i])
            strand(alns) <- "*"
            hits <- countOverlaps(alns,txdb)
            l[[i]] <- countOverlaps(txdb, alns[hits==1])
            names(l) <- gsub("\\.bam", "", files)
        }
        ct<-as.data.frame(l)
        ct
}
```

The function generateCountTable() works simply as follows. First, change the working directory to point to the same directory that contains the BAM files and then run the function with the following command:

```
counttable<-generateCountTable(dir(pattern=".bam"))
```

The result should be a count table similar to:

```
head(counttable)
```

	Gm12892_1_chr18	Gm12892_2_chr18	Gm12892_3_chr18	hESC1_chr18	hESC2_chr18	hESC3_chr18	hESC4_chr18
1	0	0	0	0	0	0	0
10	0	0	0	0	0	0	0
100	0	0	0	0	0	0	0
1000	27	72	12	4446	3300	3605	3498
10000	0	0	0	0	0	0	0
100008586	0	0	0	0	0	0	0

The resulting count table can then be used for further analyses as specified in the later chapters.

7.5 REPRESENTING GENOMES IN R

Genomes are represented by objects of Biostring type. The human genome assembly versions 17–19 are readily available from the Bioconductor project. The current genome version is contained in the package BSgenome.Hsapiens.UCSC.hg19. The genome consists of several Biostring objects, one for each chromosome, and they can be extracted as separate objects, if needed:

```
library(BSgenome.Hsapiens.UCSC.hg19)
chr18<-(BSgenome.Hsapiens.UCSC.hg19[["chr18"]])
```

One of the interesting uses of the genome data is remapping of reads. Mapping functionality in R does not fully compete with the functionality with external mappers, but it can be done. The following example shows how to perform mapping in R. First, we need to read the reads as sequences into R. This is accomplished by specifying the parameters for the function `readBamGappedAlignments()`:

```
p2 <-ScanBamParam(what=c("rname", "strand", "pos",
                          "qwidth", "seq"))
h1b<-readBamGappedAlignments("hESC1_chr18.bam",
                              param=p2)
```

The sequences need then be converted into a DNAStringSet object. This is most easily done by extracting the metadata columns from the h1b object and then further extracting the DNA sequences from it. The DNA sequences are by default stored in a suitable format. Let us only use some 1000 reads for this example:

```
seqs<-mcols(h1b)$seq
seqs2<-seqs[100:1100,]
```

The mapping can then be performed with the function `matchPDict()`:

```
mpd<-matchPDict(seqs2, chr18)
```

The mapping is by default done so that no mismatches or indels are allowed in the pairwise alignments. Once the reads are mapped, we can count how many of the mapped reads are located into the known genes. The counting principle was already covered above, but getting the mapped reads into a suitable format takes some gymnastics: first the object containing the mapped reads is converted into a CompressedIRangesList, which is then in turn converted into a GRanges object. Both conversions are done using the `as()` function. The final analyzable GRanges object is generated from this temporary GRanges object. The whole conversion carousel is detailed below:

```
mpd2<-as(mpd, "CompressedIRangesList")
mpd3<-as(RangedData(mpd2), "GRanges")
gr<-GRanges(seqnames=Rle(rep("chr18", length(mpd3))),
            ranges=mpd3@ranges,
            strand=strand(mpd3),
```

```
            seqinfo=Seqinfo("chr18", 78077248)
            )
```

Once the objects are in a suitable format, we can count the hits to the genes:

```
txdb_chr18<-keepSeqlevels(txdb, "chr18")
hits <- countOverlaps(gr, txdb_chr18)
ol<-countOverlaps(txdb, gr[hits==1])
```

In the final result, there are two genes that have reads mapped to them, one with 178 reads and another with 262 reads. A total of 22,930 genes do not have any genes mapped to them:

```
table(ol)
ol
    0  178   262
22930    1     1
```

7.6 REPRESENTING SNPs IN R

Bioconductor offers an SNPlocs-package for at least human genome. The current SNP location package at the time of writing was based on dbSNP database version 137 and is named SNPlocs.Hsapiens.dbSNP.20120608. The SNP locations can be injected into the genome sequence, if needed, using the function `injectSNPs()`. After injection, the SNP information can be taken into account during, for example, mapping of the probes and other similar operations. Injecting the SNPs into the genome is carried out simply as

```
library(SNPlocs.Hsapiens.dbSNP.20120608)
library(BSgenome.Hsapiens.UCSC.hg19)
genome <- injectSNPs(BSgenome.Hsapiens.UCSC.hg19,
                     "SNPlocs.Hsapiens.dbSNP.20120608")
```

After injection of the SNPs outlined above, the object genome can be used as a genome sequence for any downstream analysis.

7.7 FORGING NEW ANNOTATION PACKAGES

Bioconductor project offers annotation packages for many model organisms. However, if the organisms you are working with is not readily available from the Bioconductor project, but it is available in some genome browser (primarily either UCSC or Ensembl), there is a fair chance that the required annotation packages can be generated for it. The same process is needed, if you want to update the annotations for your organisms, but Bioconductor has not yet done it (there is a 6-month update cycle for Bioconductor).

Genome-wide annotation packages can be generated using the AnnotationForge package. Let us generate a new genome-wide annotation package for alpaca, which is a cute, fluffy animal. For the work, we will need the taxonomic name of the species (*Vicugna pacos*) and its taxon id at the NCBI's genome database (30538). The actual work is done using the function `makeOrgPackageFromNCBI()`:

```
library(AnnotationForge)
makeOrgPackageFromNCBI(version = "0.1",
       author = "JarnoTuimala < name@server > ",
       maintainer = "JarnoTuimala < name@server > ",
       outputDir = "C:/Users/JarnoTuimala/Desktop/
                   alpaca",
       tax_id = "30538",
       genus = "Vicugna",
       species = "pacos")
```

This will create an annotation package to the path specified with the argument outputDir (here, on Jarno Tuimala's Desktop in the folder alpaca). Once the forging has been finalized, the resulting package can be installed to R with the function `install.packages()`:

```
install.packages(pkgs="C:\\Users\\Jarno Tuimala\\
                       Desktop\\alpaca\\org.Vpacos.eg.db",
                       lib="C:\\Users\\ Jarno Tuimala\\
                       Documents\\R\\win-library\\3.0",
                       type="source", repos=NULL)
```

Once the organism-specific package is in place, we can generate a separate transcript package for the alpaca. For that you would need to know the table where the required information is stored at the UCSC's Genome database. The available tables can be listed in R with the function `supportedUCSCtables()` and the exact table name can be found out at the UCSC's site under the Table Browser functionality. For alpaca, the table is xenoRefGene. Similarly, and from the same Table Browser, we need to find out the name of the alpaca's genome at the UCSC's database. It seems to be vicPac2.

After finding out the correct table name and the name of the genome, we can produce a transcriptDb object using the function `makeTranscriptDbFromUCSC()` from the GenomicFeatures package:

```
txdb <- makeTranscriptDbFromUCSC(genome="vicPac2",
                                  tablename="xenoRefGene")
```

Using the transcriptDb object, a package can be assembled with the following command:

```
makeTxDbPackage(txdb,
      version="0.1",
      maintainer="JarnoTuimala<name@server>",
      author="JarnoTuimala<name@server>",
      destDir="C:/Users/Jarno Tuimala/Desktop/alpaca",
      license="Artistic-2.0")
```

Note that destination directory specified with the argument destDir has to exist, so you might need to create it first.

Or alternatively, if you would rather use Ensembl for constructing the transcript annotation package (and without the additional step of building the object txdb first!), you can use the command:

```
makeTxDbPackageFromBiomart(
   version="0.1",
   maintainer="Jarno Tuimala<name@server>",
   author="Jarno Tuimala<name@server>",
   destDir="C:/Users/Jarno Tuimala/Desktop/alpaca2",
   license="Artistic-2.0",
   biomart="ensembl",
   dataset="vpacos_gene_ensembl",
   transcript_ids=NULL,
   circ_seqs=DEFAULT_CIRC_SEQS,
   miRBaseBuild=NA)
```

The resulting package can be installed similarly to the organism-specific package, which was covered above.

In addition to the organism-specific package and the transcript package, a new genome package can be forged using the functions from the BSgenome software package. SNP packages are slightly trickier than others to produce, but the tools folder under the SNPlocs package contains the Linux bash scripts that can be modified to generate SNP package for human as well as other organisms. These packages are not covered in detail here, because they are less frequently in need of updating than the other annotation packages and they also require extra tinkering to get to work compared to the most common model organisms.

7.8 SUMMARY

R ships with a good set of documentation that can be accessed inside R or online, at the user's discretion. Bioconductor uses S4 OOP system rather extensively, something that is in contrast to much of the CRAN archive. In addition to the OOP implementations of functions, Bioconductor also offers a broad set of different kinds of annotations for many model organisms. If annotation packages for an organism are not available from the Bioconductor project, you might need to or want to produce such a package from scratch.

REFERENCES

1. R Core Team. *R: A Language and Environment for Statistical Computing.* Vienna, Austria: R Foundation for Statistical Computing, 2013. http://www.R-project.org/.
2. Gentleman R.C., Carey V.J., Bates D.M., Bolstad B., Dettling M., Dudoit S., Ellis B. et al. Bioconductor: Open software development for computational biology and bioinformatics. *Genome Biology*, 5:R80, 2004.
3. Chambers J.M. *Programming with Data: A Guide to the S Language.* Berlin: Springer, 1998. ISBN 0-387-98503-4.
4. Chambers J.M. *Software for Data Analysis Programming with R.* Berlin: Springer, 2008. ISBN 0-387-75935-2.

CHAPTER 8

Differential Expression Analysis

8.1 INTRODUCTION

Differential expression (DE) analysis refers to the identification of genes (or other types of genomic features, such as, transcripts or exons) that are expressed in significantly different quantities in distinct groups of samples, be it biological conditions (drug-treated vs. controls), diseased vs. healthy individuals, different tissues, different stages of development, or something else. Although genes (if we focus on those for a while) are of course not expressed independent of each other, differential expression analysis is typically done on one gene at a time (although information is sometimes borrowed across genes, as we will see below), that is, in a *univariate* way. The reason for this is that while there may be tens of thousands of genes for which expression measurements have been done, the number of biological samples is typically much smaller. Another way of stating this is that the number of *examples* is much smaller than the number of *features*, which makes it harder to fit a statistical model that considers all genes as a whole. Multivariate dimension reduction methods such as principal component analysis (PCA) [1] or nonnegative matrix factorization (NMF) [2] can be used to construct low-dimensional representations of the expression profiles that retain some of the properties of the complete data set and are thus often useful for visualization and sometimes as a preprocessing step for analysis.

DE analysis of RNA-seq data differs from microarray DE analysis in that the observed data are in the form of discrete counts generated from a sampling process, while microarray measurements are continuous measurements of a fluorescence signal. One aspect of this is that, because RNA-seq is a sampling procedure, there is a certain amount of "real estate" (the total number of all reads from the sequencing instrument) that the actual transcripts in the sequencing library have to "share." This means that highly expressed transcripts will often make up a large amount of the sequencing library, and in a shallow sequencing experiment less expressed genes may not be represented in the final data even though they were present. By contrast, microarrays are not constrained to such a "zero-sum game," although they of course have other limitations. An attractive feature of RNA-seq, though, is the possibility to re-sequence the same library to potentially recover more expressed transcripts.

8.2 TECHNICAL VERSUS BIOLOGICAL REPLICATES

Let us take a moment to think about *replication* and how it helps in differential expression analysis. The word replication can have several meanings but what we mean here is that we get more than one measurement of the quantity of interest. For example, if we want to compare tissue-specific expression in fruit fly and sequence one sample of mRNA extracted from a salivary gland and one sample from the spinal cord, we have an unreplicated experiment.

Replication is considered one of the three cornerstones of proper experimental design outlined by Fisher (1935): randomization, replication, and blocking. An excellent explanation of these concepts in the context of RNA sequencing can be found in a paper by Auer and Doerge [3], which is a highly recommended reading before you plan your experiment.

The purpose of replication is to be able to estimate the variability between and among groups, which is important for, for example, hypothesis testing. Technical replication is used to estimate the variability of the measurement technique, for example, RNA-seq. Biological replication is used to find out the variability within a biological group. Roughly speaking, a change observed in gene expression between two groups can only be called significant if the difference between the groups is large compared to the variability within the group, while taking the sample size into consideration.

There can be different kinds of technical replicates, for example, sequencing the same library in two different lanes of a sequencer or different library preparations performed on the same sample of extracted

RNA. Typically, the RNA extraction would be the same in technological replicates, but different in biological replicates. There are also borderline cases where it is hard to call replicates "biological" even if they are taken from sources that could be considered different, for instance, different cultures of the same genetically homogeneous cell line. In these cases, the important thing is to think about what questions a given differential expression comparison would actually answer, rather than getting hung up on the terminology.

How many replicates should you use? This depends on the specifics of the experiment. The biological homogeneity of the different samples, the purpose of the experiment and the desired level of statistical power, among other things, will affect the number of replicates needed. You may want to try a power calculation tool for RNA-seq such as Scotty (http://euler.bc.edu/marthlab/scotty/scotty.php) to determine the number of replicates.

Many sequencing core facilities require or suggest using at least three or four replicates per group to be compared; two is almost always too few. With three, there is the risk that at least one sample will fail in library preparation or sequencing and you still end up with only two replicates in one of the groups.

Anecdotally, human blood and some tissue samples used for clinical case–control transcriptomics studies seem to exhibit considerable variation between individuals. Particularly for complex diseases, very large numbers of replicates (perhaps hundreds or thousands) may be needed to observe differential expression between cases and controls. For cell lines or samples from distinct tissues, only a few replicates may be needed.

8.3 STATISTICAL DISTRIBUTIONS IN RNA-SEQ DATA

Expression levels of the same gene across different cells have been shown to follow a *log-normal* distribution as measured by quantitative PCR [4]. (Note that this is different from asking about the expression distribution of different genes in the same cell.) However, most gene expression experiments are done in a population of cells and a few different distributions have been proposed.

For RNA-seq experiments, where one might assume that sequences are sampled at random from the sequencing library, the raw read counts would be expected to be Poisson-distributed. If you think about it for a minute, you would expect to get slightly different counts even for the same library in an idealized scenario where it was sequenced twice under the same conditions. This inevitable noise which arises from the sampling process is called *shot noise*, and often the variability between technical replicates in RNA-seq can be described quite well by this type of Poisson noise. However, when samples are taken from biologically distinct sources, such as different individuals, the variability between them has often been modeled by a *negative binomial* distribution (sometimes called gamma-Poisson distribution). This distribution can be described as an *overdispersed* Poisson distribution—a version of that distribution but with higher variance. While a Poisson distribution has the same variance as its mean μ, the negative binomial distribution's variance can be written as $\sigma^2 = \mu + (1/r)\mu^2$ where r is a positive integer (which means that the variance will always be larger than the mean). A number of popular packages such as *DESeq* [5] and *edgeR* [6] use the negative binomial distribution as the basis of their modeling of RNA-seq counts.

However, RNA-seq count data also show some characteristics like *zero inflation* (a large proportion of values with zero counts) which makes it harder to fit a negative binomial distribution. A recent paper [7] argues that RNA-seq count profiles can be modeled using a more general family of distributions, the Poisson–Tweedie family, and the authors have provided an R package, *tweeDESeq*, implementing this approach.

The *limma* [8] package, which has long been used for microarray analysis, takes another approach and first transforms the raw count data (using the *voom* function) into continuous values with associated confidence weights and then proceeds to use the statistical framework developed for microarrays on these values. The DESeq2 package [9], an updated version of DESeq, can also implement similar transformations.

Non-parametric methods such as *SAMSeq* [10] and *NOISeq* [11] do not assume anything about the form of the distribution but rather rank the genes by expression and use statistics and tests based on these ranked lists, and random permutations of those lists, to identify differentially expressed genes (Table 8.1).

8.3.1 Biological Replication, Count Distributions, and Choice of Software

The number of biological replicates you have access to can affect the choice of differential expression analysis software. With a fair number of biological replicates (perhaps at least 5–10 biological replicates per group, depending on the specifics of the data set), it may be beneficial to use a nonparametric method that does not make assumptions about the form of the statistical distribution of the observed data. For the far more common scenario with few

TABLE 8.1 List of (Some) Software Tools for Differential Expression Analysis

Software Tool	Type of Software	Analysis Approach	Comment
DESeq [5]	R/Bioconductor package	Count-based (negative binomial)	Considered conservative (low false-positive rate)
DESeq2 [9]	R/Bioconductor package	Count-based (negative binomial)	Recommended over DESeq by authors; less conservative than DESeq
edgeR [6]	R/Bioconductor package	Count-based (negative binomial)	Similar to DESeq in philosophy
tweeDESeq [7]	R/Bioconductor package	Count-based (Tweedie distribution family)	More general than DESeq/edgeR, but new and not widely tested
Limma [8]	R/Bioconductor package	Linear models on continuous data	Originally developed for microarray analysis, very thoroughly tested. Need to preprocess counts to continuous values
SAMSeq [10] (samr)	R package	Nonparametric test	Adapted from the SAM microarray DE analysis approach. Works better with more replicates
NOISeq [11]	R/Bioconductor package	Nonparametric test	
CuffDiff [18]	Linux command line tool	Isoform deconvolution + count-based tests	Can give differentially expressed isoforms as well as genes (also differential usage of TSS, splice sites)
BitSeq [21]	Linux command line tool and R package	Isoform deconvolution in a Bayesian framework	Can give differentially expressed isoforms. Also calculates (gene and isoform) expression estimates
ebSeq [22]	R/BioConductor package	Isoform deconvolution in a Bayesian framework	Can give differentially expressed isoforms. Can be used in a pipeline preceded by RSEM expression estimation

biological replicates, nonparametric methods are typically underpowered. In these cases, it makes sense to use parametric methods that assume certain form of the distribution based on empirical data, like the above-mentioned DESeq and edgeR packages (which use the negative binomial distribution) or tweeDESeq (which uses the Poisson–Tweedie distribution family). Recent reports have also shown limma to perform well in such scenarios.

8.4 NORMALIZATION

Typically, RNA-seq data are normalized in some way ahead of or as a part of differential expression analysis (most packages only ask for raw counts and perform the normalization internally). This could be for a couple of reasons:

- To enable comparisons between samples
- To enable comparisons between genes
- To make the expression level distribution conform to assumptions used in statistical methods

The standard RPKM measure (or its paired-end equivalent, FPKM) was introduced in a 2008 paper [12] and designed to enable comparisons of the same gene's expression levels across samples or of different genes' expression levels in the same sample. RPKM (FPKM) stands for Reads (Fragments) Per Kilobase per Million (mapped) reads and it corrects the raw counts both with respect to the gene or transcript length and sequencing depth. There are some slight variations in how the measure has been applied; for instance, different studies have used different values in the denumerator: the total number of reads from the sequencer, the number of reads mapped to the genome or transcriptome, or the number of reads mapped to known exons. Similarly, for the gene or transcript length, some investigators have chosen to use the entire length of the transcript, while others have used the "effective length" [13] or the "mappable length" [14].

While R/FPKM is still the most commonly used measure for expression from RNA-seq, other measures have been proposed to correct for possible biases in some scenarios. TPM (transcripts per million) [15] is very similar to R/FPKM but also accounts for the distribution of transcript lengths in the RNA population. Without this correction (as in R/FPKM), the authors argue that bias will be introduced when comparing two RNA pools with different transcript length distributions. TMM (trimmed means of *M* values) also attempts, in another way, to correct for different compositions of RNA pools. The following thought experiment from [16] explains the idea:

> Imagine we have a sequencing experiment comparing two RNA populations, A and B. In this hypothetical scenario, suppose every gene that is expressed in B is expressed in A with the same number

of transcripts. However, assume that sample A also contains a set of genes equal in number and expression that are not expressed in B. Thus, sample A has twice as many total expressed genes as sample B, that is, its RNA production is twice the size of sample B. Suppose that each sample is then sequenced to the same depth. Without any additional adjustment, a gene expressed in both samples will have, on average, half the number of reads from sample A, since the reads are spread over twice as many genes. Therefore, the correct normalization would adjust sample A by a factor of 2.

An important distinction between TMM, on the one hand, and TPM and R/FPKM, on the other hand, is that TMM is a batch normalization method, that is, it is not designed for use on a single sample, but on a group of samples. Thus, the correction factors from TMM normalization should be recalculated each time the set of samples changes, while TPM and R/FPKM normalizations are local to the sample and not affected by other samples. Another distinction is that TMM does not consider transcript length. If you are doing standard differential expression analysis, however, this does not matter, because you are not comparing different transcripts to each other, but rather the same transcript across conditions, so the transcript length is always the same.

A useful review of RNA-seq normalization methods was provided by Dillies et al. [17].

FACT BOX: HOW TO SELECT A SOFTWARE PACKAGE FOR DIFFERENTIAL EXPRESSION ANALYSIS

Here is a simple decision tree you might use to pick a software package depending on your needs.

Select type of feature to test differential expression for

- Differentially expressed **exons** => *DEXSeq*
- Differentially expressed **isoforms** => *BitSeq, Cuffdiff or ebSeq*
- Differentially expressed genes => **Select type of experimental design**
  - Complex design (more than one varying factor) => *DESeq, edgeR, limma*
  - Simple comparison of groups => **How many biological replicates?**
    - More than about 5 biological replicates per group => *SAMSeq*
    - Less than 5 biological replicates per group => *DESeq, edgeR, limma*

8.5 SOFTWARE USAGE EXAMPLES

We will look at examples of how to use two popular programs for differential expression analysis, Cuffdiff and DESeq, each representing one of the typical workflows in differential expression analysis (see Figure 8.1).

8.5.1 Using Cuffdiff

The Cuffdiff program is part of the popular Cufflinks package for RNA-seq assembly, quantification, and differential expression analysis. It can assess differential expression both on the gene and transcript levels simultaneously. The authors of the package argue that Cuffdiff is better than common alternatives like DESeq and edgeR in that it combines the deconvolution of expression data into isoforms with differential expression testing [18]. By contrast, DESeq and edgeR typically take gene counts from a software package like *HTSeq*, *BEDTools*, or *featureCounts*, thereby not considering isoforms, which will lead to bias in differential expression

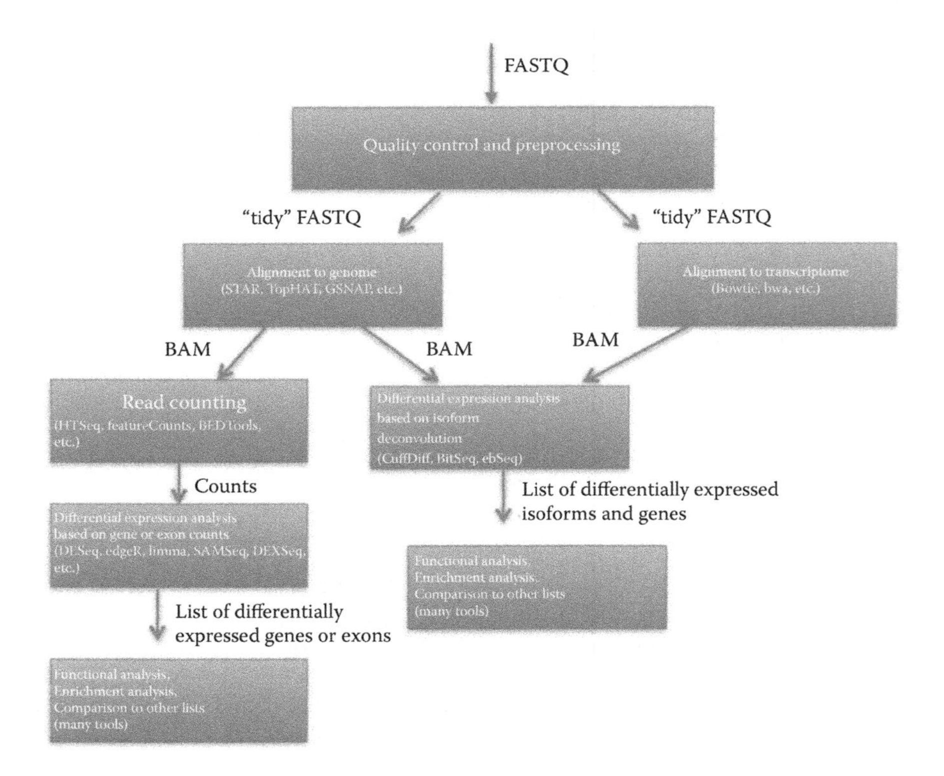

FIGURE 8.1 Typical workflows used in RNA-seq differential expression analysis.

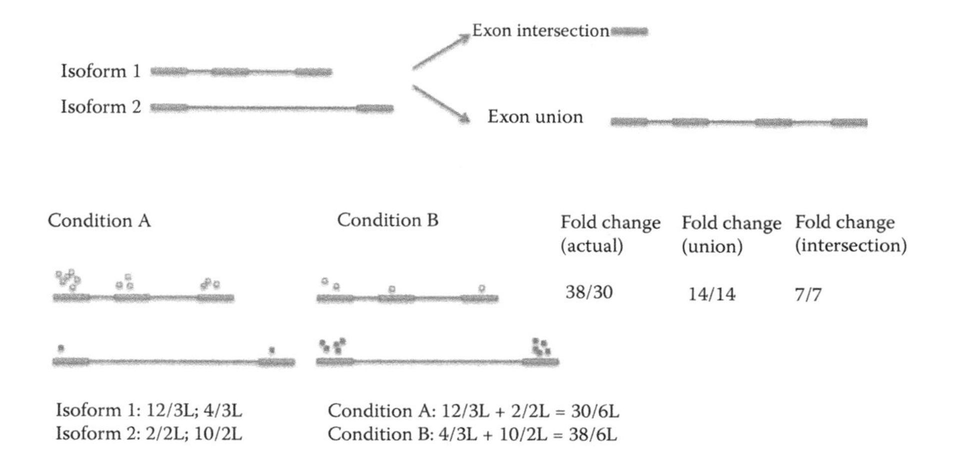

FIGURE 8.2 Isoforms need to be considered in order to obtain unbiased gene-level expression estimates. An imaginary gene with two different isoforms is depicted above (upper left corner). For simplicity, all exons are assumed to have the same length L. Two common methods for calculating gene-level counts is the exon-intersection model, where only reads mapping to exons that are part of all of the gene isoforms are considered, and the exon-union model, where all reads mapping to any exon are considered (upper right corner). In the hypothetical case shown in the lower panel, where the reads originating from each isoform in two different conditions A and B are indicated, the actual fold change for the whole gene considering the isoforms would be estimated as (38/30), while both the exon-intersection model and the union model would estimate it as 1 (i.e. no change at all).

calculations on the whole-gene level whenever more than one isoform is expressed (Figure 8.2).

However, a couple of comparative studies of RNA-seq DE software (for instance, see [19]) have shown that this advantage of Cuffdiff does not necessarily lead to better results in practice. Another advantage of Cuffdiff is that its output can be fed into a very useful visualization and analysis package in R, *cummeRbund* [20]. On the other hand, Cuffdiff does not support more complex experimental designs in the way that DESeq, edgeR, and limma do. In this context, we should also mention *BitSeq* [21] and *ebSeq* [22], which share the advantages and limitations of Cuffdiff.

FACT BOX: PROS AND CONS OF CUFFDIFF

Pros

- Computes isoform expression levels
- Tests for differential expression of genes, isoforms, splice sites, transcription start sites
- Good visualization support

Cons

- No support for factorial designs (can only compare two groups)

Let us try to run Cuffdiff on our example data. Cuffdiff accepts SAM/BAM alignment files as input and it also needs a transcript annotation file in GTF format to define the set of genomic features that will be assessed.

In the command below, we are only using some of the many options in Cuffdiff(2). You can see all of them by running the program without input arguments. The `–o` option specifies the output directory, which will usually contain many files and which is the structure to be inputted as a whole into cummeRbund for visualization and analysis if one so desires. `–p 4` specifies that we would like to use four processors, `–L` lists the labels we want to use for the "conditions" (in our case, these are rather cell types: hESC and GM12892), `--FDR` gives the false discovery rate cutoff for the DE analysis, and `–u` specifies that we want to use something called "multi-read correction" which is generally recommended. Finally, the mandatory input arguments are the GTF annotation files (note that this must have the same chromosome or contig names as the BAM/SAM files!) and the comma-separated lists of BAM files for the same group. There is a whitespace between the BAM files belonging to each group to indicate that fact, and a comma between each file name in each group to indicate that they are replicates. Assuming that $CUFFPATH is an environment variable containing the path to the directory where the cuffdiff executable is located, and that $GTFPATH similarly points to a directory containing a GTF file, the commands could look as follows.

```
$CUFFPATH/cuffdiff -o chr18_hESC_vs_GM12892 -p 4 -L
hESC,GM12892 --FDR 0.01 -u $GTFPATH/Homo_sapiens.
GRCh37.59.chr-added.gtf hESC1_chr18.bam,hESC2_chr18.
bam,hESC3_chr18.bam,hESC4_chr18.bam Gm12892_1_chr18.
bam,Gm12892_2_chr18.bam,Gm12892_3_chr18.bam
```

We may get some warnings about fragment lengths but will ignore those here. After the command has finished, we find a couple of files with differential expression information in the specified output folder. The `gene_exp.diff` file contains the gene-level information. The first three lines look like this:

```
test_id gene_id gene     locus   sample_1         sample_2        status  value_1 value_2 log2(fold_change)        test_stat        p_value q_value significant
ENSG00000000003 ENSG00000000003 TSPAN6  chrX:99883666-99894988  hESC    GM12892 NOTEST  0       0       0        0       1        1       no
ENSG00000000005 ENSG00000000005 TNMD    chrX:99839798-99854882  hESC    GM12892 NOTEST  0       0       0        0       1        1       no
```

These are not very interesting as these first few genes (the genes are always ordered by the gene_id column) are not on chromosome 18, and since we have no data outside of that chromosome in our BAM files, Cuffdiff has not even checked for differential expression here, hence the NOTEST value in the seventh ("status") column.

We can see that 237 genes on chromosome 18 have been called as differentially expressed from looking at rows with "yes" in the last column (meaning that the gene is significantly differentially expressed between the two groups). One way to find this out is to use the UNIX awk command:

```
awk'$14=="yes"' gene_exp.diff|wc-l
```

A few of the lines corresponding to such genes might look like this (note the "yes" in the last column).

```
ENSG00000017797 ENSG00000017797 RALBP1  chr18:9475006-9538112    hESC    GM12892 OK      0        5453.82 inf     nan     5e-05   6.07477e-05     yes
ENSG00000039139 ENSG00000039139 DNAH5   chr5:13690439-13944652   hESC    GM12892 OK      207.984 0        -inf    nan     5e-05   6.07477e-05     yes
ENSG00000040731 ENSG00000040731 CDH10   chr5:24487208-24645087   hESC    GM12892 OK      238.193 0        -inf    nan     5e-05   6.07477e-05     yes
```

The first six columns display gene IDs, common names, chromosomal coordinates, and sample group labels, respectively. The status (seventh) column now has the value OK, which means that Cuffdiff has had enough data to perform a significance test for the gene in question. The following columns (8 and 9) contain the mean FPKM in each group, hESC or GM12892, respectively. The tenth column indicates the log 2 fold change between the groups' mean FPKMs; since the mean was zero in one of the groups in these particular cases, the log fold change is infinite. The 11th column contains the value of a test statistic, here "nan" (not a number), which is related to the infinite log fold change. The 12th column has the estimated p-value, and the 13th column has the estimated q-value (p-value corrected for multiple comparisons). Usually, the most interesting columns are those containing the q-value and log 2 fold change.

Correspondingly, the `isoform_exp.diff` file contains similar information on the isoform level. There are 238 differentially expressed isoforms—unusually close to the gene-level value.

8.5.2 Using Bioconductor Packages: DESeq, edgeR, limma

There are many differential expression analysis packages available for RNA-seq in R's BioConductor project. Here, we will mainly focus on DESeq2 but also discuss edgeR and limma packages, partly because they are popular and partly because they currently have the best support for factorial designs. Other useful packages are the nonparametric NOISeq, the Bayesian methods BitSeq, baySeq, and ebSeq, and tweeDESeq.

8.5.3 Linear Models, the Design Matrix, and the Contrast Matrix

The DESeq(2), edgeR, and limma packages are all based on the concept of (generalized) *linear models* (in fact, limma stands for "linear models for microarray data"). It is beyond the scope of this book to provide a proper introduction to this subject, so we urge the reader to consult a good statistics textbook. The basic idea is to model the expression of each gene as a linear combination of some different explanatory variables (or *factors*). For instance, if you have an experiment involving different patients, treatments, and time points, the linear model for each gene could be thought of as

$$y = a + b \cdot \text{treatment} + c \cdot \text{time} + d \cdot \text{patient} + e$$

where y is the gene expression measured in some unit, e an error term, and a, b, c, d are parameters to be estimated from the data. a is called the *intercept* and represents the average expression level of the gene when all the other factors (treatment, time, and patient) are in their reference state. (You can choose any combination of these as the reference state; it is arbitrary.)

A *generalized* linear model (GLM) is a more flexible version of a standard linear model which, among other things, allows the distribution of the response variable to be different from the normal distribution used in standard linear regression. The GLMs used in edgeR and DESeq assume that the read counts are distributed according to the negative binomial distribution.

Another feature of DESeq, edgeR, and limma is that these methods can borrow information across genes to increase statistical power. The packages use different schemes to achieve *moderated* variance estimates where each gene's variance is modeled as a weighted combination of the gene's own variance estimated from the gene-specific data and the average variance of all genes, or a subset of the genes.

More generally, the linear model may be written in matrix form as

$$y = X \cdot \beta + \varepsilon$$

where y is again the expression level and ε is an error term. β is a vector of parameters to be estimated from the data, and X, which describes the experimental factors involved, is called the *design matrix*.

The design matrix as well as the *contrast matrix* are two concepts that you should familiarize yourself with before attempting an analysis with DESeq, edgeR, or limma, although you may not have to use them directly—for instance, they are dealt with implicitly in DESeq2. These concepts are commonly used in experimental design and we recommend that you look them up in a good statistics textbook. The limma user guide [23] also contains useful examples of how to set up experimental and design matrices.

8.5.3.1 Design Matrix

When we talk about the design matrix in the following, we will refer to the R object that describes the design of your experiment (although it is actually a more general concept, as mentioned above). As an example, if you are analyzing an experiment where tumor and healthy tissue samples have been taken from the same patients, you might have a table `expTable` describing the experiment specified like this:

```
expTable <- data.frame(Individual=c("Patient1","Pati
  ent1", "Patient2","Patient2"),Status=c("Tumor","Heal
  thy", "Tumor","Healthy"),row.
  names=paste0("sample",1:4))
```

```
expTable
          Individual   Status
sample1     Patient1    Tumor
sample2     Patient1  Healthy
sample3     Patient2    Tumor
sample4     Patient2  Healthy
```

This could be turned into a design matrix with the `model.matrix()` function:

```
design.matrix <- model.matrix (~Individual+Status,
  data = expTable)
```

The design matrix may look a bit cryptic but is not hard to interpret after you have had some practice.

```
>design.matrix
        (Intercept)   IndividualPatient2    StatusTumor
sample1           1                    0              1
sample2           1                    0              0
sample3           1                    1              1
sample4           1                    1              0
attr(,"assign")
[1]0 1 2
attr(,"contrasts")
attr(,"contrasts")$Individual
[1]"contr.treatment"

attr(,"contrasts")$Status
[1]"contr.treatment"
```

You will note that there are three columns: (Intercept), Individual Patient2, and StatusTumor. These columns contain binary *indicator variables*, which indicate whether a certain factor has a certain value in a sample with a 1 (and contain a 0 otherwise). The intercept column contains the same value, 1, for all the samples, and simply indicates that the linear model which will be built for each gene will have an intercept term. The intercept corresponds to the average expression level when all the experimental factors are in their reference state, which in this case would be (Individual = Patient1) and (Status = Normal). The second column in the design matrix, IndividualPatient2, indicates the samples taken from Patient 2 (so the Individual factor has the value Patient2) with a 1 on the corresponding row and a 0 otherwise. The third column, StatusTumor, has a 1 for the samples where the Status factor has the value Tumor.

8.5.3.2 Contrast Matrix

When you actually want to test for differential expression, you may need to set up a *contrast matrix* that describes which comparisons you want to make (this is not necessary in DESeq2). This matrix often has just one nonzero element (if you are doing a single comparison). For example, with the design matrix given above, you might specify a comparison between tumor and normal tissue using the `makeContrasts` function from the limma package:

```
contrast.matrix<- makeContrasts(StatusTumor,
  levels=design.matrix)
```

If you wanted to compare Patient 1 and Patient 2, you could specify the contrast as

```
contrast.matrix<- makeContrasts(IndividualPatient2,
  levels=design.matrix)
```

Since there is an intercept term in the model, there is no column for (Individual = Patient 1), which is regarded as the reference value of the Individual factor. "IndividualPatient2" therefore implicitly describes the difference between Patient 2 and Patient 1.

Alternatively, by using the ~0 notation in the `model.matrix` function, the intercept is omitted from the model and the design matrix will not have the (Intercept) column:

```
design.matrix<- model.matrix (~0+Individual+Status,
  data = expTable)
design.matrix
   IndividualPatient1 IndividualPatient2  StatusTumor
sample1             1                  0            1
sample2             1                  0            0
sample3             0                  1            1
sample4             0                  1            0
attr(,"assign")
[1]1 1 2
attr(,"contrasts")
attr(,"contrasts")$Individual
[1]"contr.treatment"

attr(,"contrasts")$Status
[1]"contr.treatment"
```

Here, there is no "reference state" as in the example above, and to get a comparison between Patient 1 and Patient 2 you would need to do

```
contrast.matrix<- makeContrasts(IndividualPatient2-
  IndividualPatient1,levels = design.matrix)
```

Please refer to the limma or edgeR tutorial or a statistics textbook for more info on design and contrast matrices.

8.5.4 Preparations Ahead of Differential Expression Analysis

One typically starts the analysis by loading a count table. Let's try that for our example data set. First, we will explain how to produce a count table starting from BAM files or individual count files if you do not have one already.

8.5.4.1 Starting from BAM Files

If you are starting from aligned files, you need to obtain counts by a program such as HTSeq, BEDTools, or featureCounts (which can be used in R via the *Rsubread* package). For HTSeq and featureCounts, you need to convert the binary BAM files into SAM files first (BEDTools can handle BAM files directly). This can be done using the *samtools* set of (command-line) tools.

```
samtools sort -no Gm12892_1_chr18.bam Gm12892_1_chr18_
sorted |samtools view - > Gm12892_1_chr18.sam
```

We use the -n option of samtools sort because we want to make sure that the SAM file is sorted on read identifier, as required by HTSeq for paired-end reads, although the requirement has been removed in the latest version.

Apply similar commands to the other samples. We can then apply HTSeq, running it as a Python module:

```
python -m HTSeq.scripts.count -s no Gm12892_1_chr18.
sam Homo_sapiens.GRCh37.70.chr18.chr.gtf > gm1.txt
```

(and similar commands for the other samples). This will generate count files for each sample.

8.5.4.2 Starting from Individual Count Files

The files provided together with the book include count files from HTSeq called `gm1.txt, gm2.txt,..., h1.txt,..., h4.txt`. There are many ways to combine these into a table, including the UNIX `join` command and various custom scripts such as the DESeqDataSetFromHTSeqCount command in DESeq2. One way to do it in R is by using commands such as

```
samples <- c(paste0("gm",1:3),paste0("h",1:4))
first.sample <- read.delim(paste0(samples[1],".
   txt"),header=F,row.names=1)
count.table <- data.frame(first.sample)
for(s in samples[2:length(samples)]){
      fname <- paste0(s,".txt")
      column <- read.delim(fname,header=F,row.names=1)
      count.table <- cbind(count.table,s = column)
}
```

```
colnames(count.table) <- samples
write.table(count.table,file="count_table_chr18.
  txt",sep="\t",quote=F)
```

8.5.4.3 Starting from an Existing Count Table

If you have already prepared a count table, you can simply load it with a command similar to

```
d.raw<- read.delim("count_table_chr18.txt",sep="\t")
```

This would load a tab-separated file; if you have a space-separated file, you would instead put `sep = "  "`, and so on.

8.5.4.4 Independent Filtering

It is typically recommended to filter out lowly expressed transcripts ahead of count-based differential expression analysis [24]. Let us require that the number of counts be more than 3 in more than two of the samples. These cutoffs are of course more or less arbitrary. In DESeq2 this step is now done automatically with the optimal cutoff being calculated by the software.

```
d<- d.raw[rowSums(d.raw>3)>2,]
```

This leaves 365 (out of 774) genes for further analysis.

8.5.5 Code Example for DESeq(2)

Now let us go on to the DESeq analysis. We will also give brief code examples for other popular packages. For a fuller explanation of the specifics of each package, please refer to the R vignette file for the package in question. These vignettes often contain useful information on package usage and design.

Load the DESeq2 package, define the relevant groups, and prepare a data frame for the group information.

```
library(DESeq2)
grp<- c(rep("GM",3),rep("hES",4))
cData<- data.frame(celltype=as.factor(grp))
rownames(cData) <- colnames(d)
```

Then use the `DESeqDataSetFromMatrix` function to construct a DESeqDataSet object. Please refer to the DESeq2 reference manual for other ways to construct such an object.

```
d.deseq<- DESeqDataSetFromMatrix(countData=d,
  colData=cData,design=~celltype)
```

The next function call actually accomplishes several distinct analysis steps which were separate commands in the original DESeq package but have been wrapped into one command for convenience in DESeq2.

```
d.deseq<- DESeq(d.deseq)
```

Now we should already have results, which can be interrogated by using the (surprise!) `results()` command. If you are wondering how we can tell that the result column is called "celltype_hES_vs_GM," the answer is that you can list the available results by calling the `resultsNames()` function.

```
res<- results(d.deseq,"celltype_hES_vs_GM")
```

Let us decide to focus on the genes for which the adjusted *p*-value is below 0.01.

```
sig<- res[which(res$padj<0.01),]
```

Keep a list of only the names of the these genes:

```
sig.deseq<- rownames(sig)
```

8.5.6 Visualization

We might want to look at the data visually in various ways. One common thing to check for potential outliers or sample mix-ups is to plot a correlation heatmap or principal component (PCA) plot of the samples. This could, of course, be done ahead of the differential expression analysis itself. However, DESeq provides a convenient function for transforming the data into a form which is more suitable for heatmap and PCA visualization:

```
vsd<- getVarianceStabilizedData(d.deseq)
```

Now we can look at the correlations between the samples:

```
heatmap(cor(vsd),cexCol=0.75,cexRow=0.75)
```

The `cexCol` and `cexRow` arguments are for making the sample labels small enough to fit in the plot (Figure 8.3).

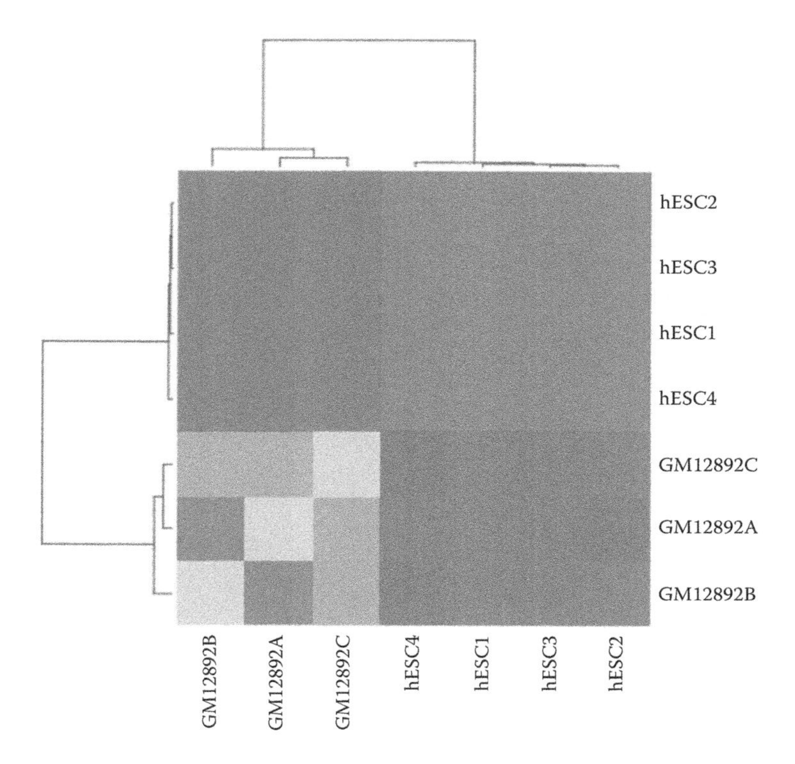

FIGURE 8.3 Correlation heatmap of cell line samples. There is an obvious grouping of samples into two distinct groups.

We can look at a PCA plot by using the `prcomp()` function in R (Figure 8.4). We need to transpose the expression matrix to get the principal component scores in terms of samples rather than genes:

```
pr<- prcomp(t(vsd))
plot(pr$x,col="white",main="PC plot",
xlim=c(-22,15))
text(pr$x[,1],pr$x[,2],labels=colnames(vsd),
cex=0.7)
```

We can also use the `biplot` function to get a similar PCA plot but which also includes information on how each gene contributes to the principal components:

```
biplot(pr,cex=c(1,0.5),main="Biplot",
col=c("black","grey"))
```

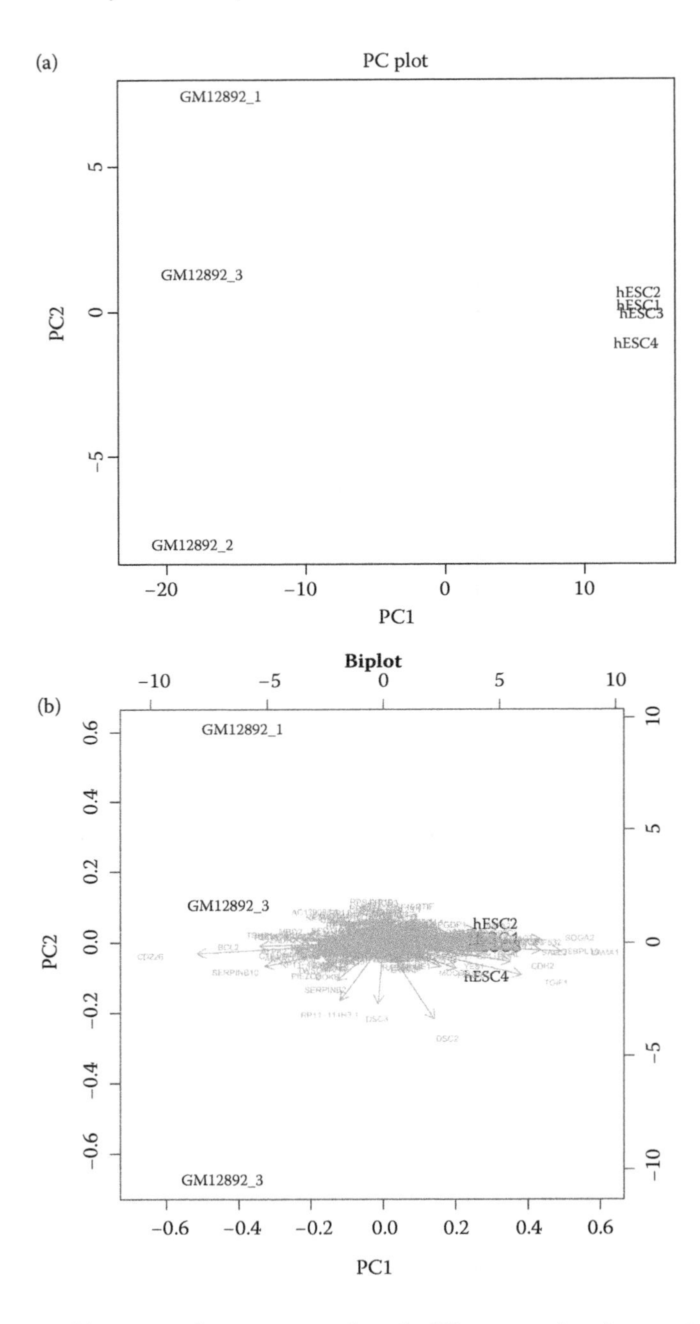

FIGURE 8.4 (a) Principal component plot of cell line samples. A grouping of the samples into two groups is evident along principal component 1 (the X axis). (b) A biplot shows both the relative locations of the samples in the PC1–PC2 space and the contributions of various genes to the principal components.

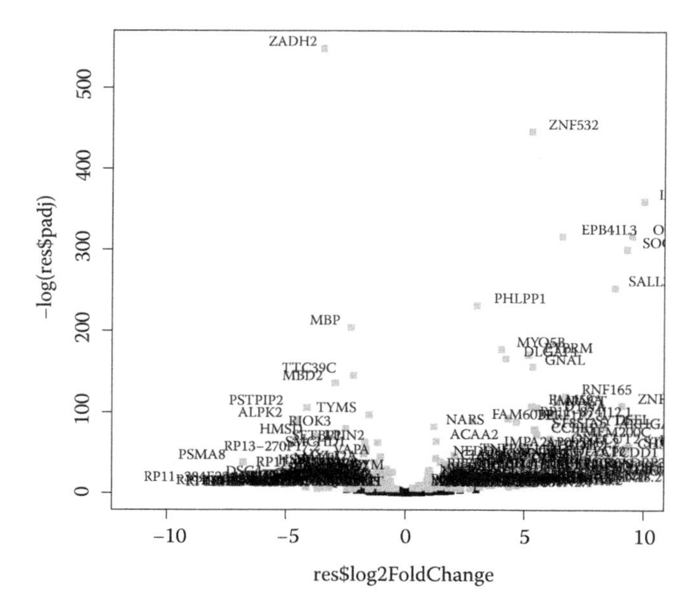

FIGURE 8.5 A volcano plot showing the negative log p value against the log fold change for each gene. Genes with adjusted p value below 0.01 are colored red.

There are a couple of different types of plots that can be useful after the differential expression analysis has been completed. A so-called volcano plot shows the (negative logarithm of the) *p*-value against the fold change for each gene, which often results in a shape like an erupting volcano. We highlight the genes with an adjusted *p*-value below 0.01 in red.

```
plot(res$log2FoldChange,-log(res$padj),pch=15)
points(sig$log2FoldChange,-log(sig$padj),
  col="grey",pch=15)
library("calibrate")# if not installed,run 'install.
  packages("calibrate")'
textxy(sig$log2FoldChange,-log(sig$padj),rownames(sig),
  cex=0.9)
```

In this case, almost all genes look differentially expressed (it is actually about half, but most of the nondifferentially expressed are squeezed together around the origin). This is perhaps not surprising as these cell lines are so different (Figure 8.5).

Finally, it might be useful to visually inspect some individual genes. Box or bar plots are often useful for this. Let us loop through to 10 most differentially expressed genes and look at box plots. First, order the genes by adjusted *p*-value:

```
sig.ordered<- sig[order(sig$padj),]
for(gene in head(rownames(sig.ordered))) {
boxplot(vsd[gene,which(grp=="GM")],vsd[gene,
which(grp=="hES")],main=paste (gene,signif(sig
[gene,"padj"],2)),names=c("GM","hES"))
readline()
}
```

Or for bar plots:

```
for(gene in head(rownames(sig.ordered))) {
barplot(vsd[gene,],las=2,col=as.numeric(as.
factor(grp)),main=gene,cex.names=0.9)
readline()
}
```

Example plots for a single gene are given in Figure 8.6.

8.5.7 For Reference: Code Examples for Other Bioconductor Packages

Example code for performing differential expression analysis using the limma, edgeR, and sam rBioConductor packages is provided below. The consistency between the different methods on our simple example scenario (GM vs. hES cells) is summarized in Figure 8.7. Reassuringly, 159 genes are identified by all four programs as being differentially expressed. (Note that the specific numbers may depend on the software version used.) DESeq, limma, and edgeR each identify only one gene that is not found by any other program, while SAMSeq identifies 42 such genes. We cannot say without additional information whether this is due to a higher sensitivity or a higher false discovery rate on SAMSeq's part; we can only say that SAMSeq seems to use more relaxed criteria for differential expression testing in this particular scenario. The level of consistency between methods achieved here is by no means typical for all experiments, and it likely arises from the very marked biological differences in the cell lines that were compared. For more similar sample groups (as in case—control studies of patients), it is more common to observe considerable inconsistency between differential expression analysis methods.

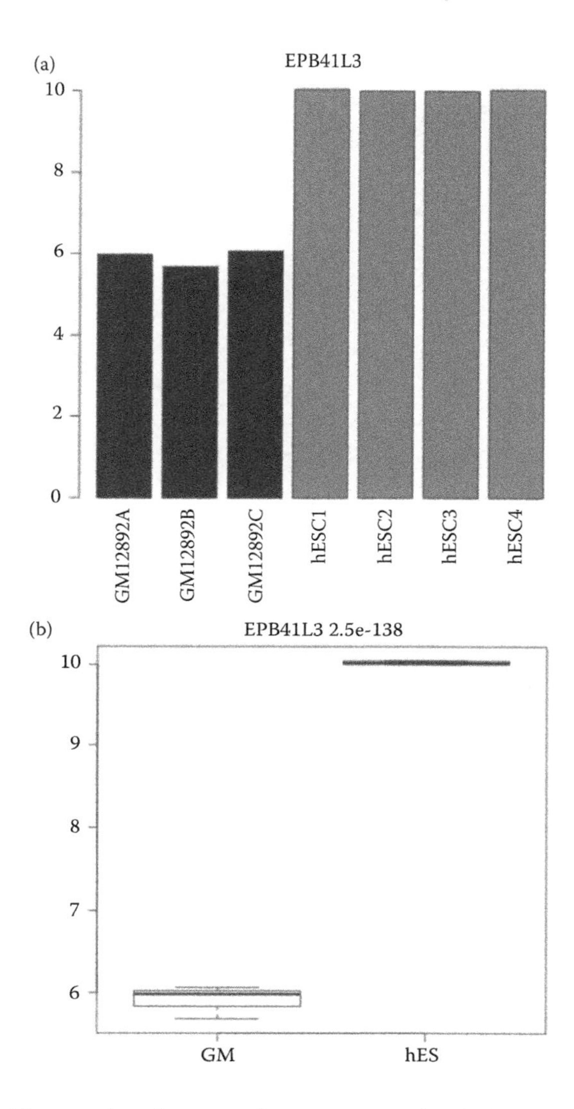

FIGURE 8.6 (a) Bar plot showing the expression level (in normalized count units) of a specific gene in the GM and hES samples. (b) Box plot showing the expression distribution of the same gene within each group (GM or hES). The bold line indicates the median.

8.5.8 Limma

```
library(limma)
grp<- c("GM","GM","GM","hES","hES","hES","hES")
des<- model.matrix(~0+grp)
```

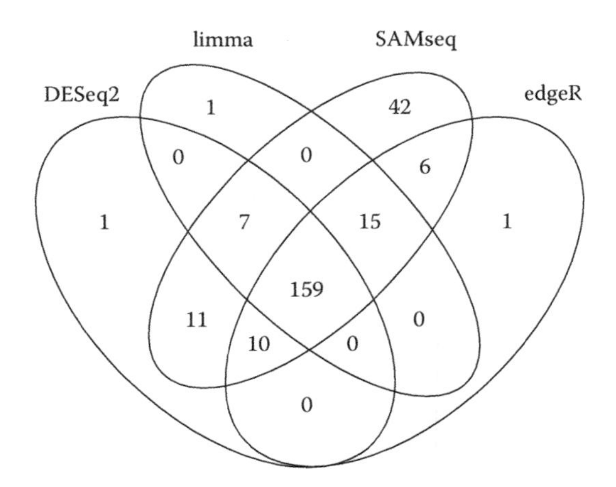

FIGURE 8.7 Consistency of differential expression calls DESeq2, limma, SAMseq and edgeR shown as a four-way Venn diagram. 159 of the genes are called as differentially expressed by all four packages.

```
colnames(des) <- c("GM","hES")
contrast.matrix<- makeContrasts(GM-hES,levels=des)

d.norm<- voom(d,design=des)

fit<- lmFit(d.norm,des)
fit2<- contrasts.fit(fit,contrast.matrix)
fit2<- eBayes(fit2)

topTable(fit2,adjust="BH")

all<- topTable(fit2,adjust.method="BH",number=10000)
sig<- all[all$adj.P.Val<1e-2,]

sig.limma<- sig$ID # If this doesn't work, try sig.
limma <- rownames(sig)
# If you want to compare the consistency of the limma
and DESeq results:
intersect(sig.limma,sig.deseq)
```

8.5.9 SAMSeq (samr package)

SAMSeq differs from the other packages in that it can give different results for the same comparison as it is based on permutation tests where it draws random subsamples from the data set. It can be useful to explicitly specify a random seed, as we have done in the code sample, in order to reproduce results.

```
library(samr)

# Note that samr is (as of Jan 2014)not in
BioConductor, so it needs to be installed using
install.packages(). However,it depends on the impute
package which is a BioConductor package.

num.grp<- c(rep(1,3),rep(2,4))
samfit<- SAMseq(d,num.grp,resp.type="Twoclass
  unpaired", genenames=rownames(d),random.
  seed=101010,fdr.output=0.01,nperms=1000)

sig.sam<- c(samfit$siggenes.table$genes.up[,1],
  samfit$siggenes.table$genes.lo[,1])
```

8.5.10 edgeR

```
library(edgeR)
edgeR.dgelist = DGEList(counts = d,group = factor(grp))
edgeR.dgelist = calcNormFactors(edgeR.
  dgelist,method = "TMM")
edgeR.dgelist = estimateCommonDisp(edgeR.dgelist)
edgeR.dgelist = estimateTagwiseDisp(edgeR.
  dgelist,trend = "movingave")
edgeR.test = exactTest(edgeR.dgelist)
edgeR.pvalues = edgeR.test$table$PValue
edgeR.adjpvalues = p.adjust(edgeR.pvalues,method = "BH")

sig.table<- edgeR.test$table[which(edgeR.
adjpvalues < 0.01),]
sig.edgeR<- rownames(sig.table)
```

8.5.11 DESeq2 Code Example for a Multifactorial Experiment

For our example of a multifactor experiment, we will focus on the differential expression testing and not spend time on preparing the data. Instead, let us install and load the *parathyroid* package, which contains read counts on the gene and exon levels for an experiment where four parathyroid adenocarcinomas were cultured and treated with two different drugs (or, as a control, untreated) at two different time points. The 24 samples (4 patients × 3 treatments × 2 time points) were subjected to RNA sequencing. One of the samples was excluded due to problems in library preparation.

```
source("http://bioconductor.org/biocLite.R")
biocLite("parathyroid")
library(parathyroid)
# You may need to install the bitops and/or DEXSeq
packages for this to work.
```

The help system (type `??parathyroid`) tells us that we can load an object containing gene-level information with the following command:

```
data("parathyroidGenes")
```

We can get information about the samples using *pData()*:

```
meta <- pData(parathyroidGenes)
head(meta)
```

	size Factor	experiment	patient	treatment	time	submission	study	sample	run
SRR479052	NA	SRX140503	1	Control	24h	SRA051611	SRP012167	SRS308865	SRR479052
SRR479053	NA	SRX140504	1	Control	48h	SRA051611	SRP012167	SRS308866	SRR479053
SRR479054	NA	SRX140505	1	DPN	24h	SRA051611	SRP012167	SRS308867	SRR479054
SRR479055	NA	SRX140506	1	DPN	48h	SRA051611	SRP012167	SRS308868	SRR479055
SRR479056	NA	SRX140507	1	OHT	24h	SRA051611	SRP012167	SRS308869	SRR479056
SRR479057	NA	SRX140508	1	OHT	48h	SRA051611	SRP012167	SRS308870	SRR479057

The `counts` command returns a count table:

```
dim(counts(parathyroidGenes))
```

If we look closely at the table with sample information, we may note that we have two entries for some samples (e.g., rows 9 and 10). These correspond to different rounds of sequencing for the same sample. This inconvenience could be handled in several ways. One way would be to simply add all counts belonging to the same sample. This would, however, mask potential bias from sequencing batches. We will instead keep the entries separate but include a term for the sequencing run into the linear model. In this way, we can attempt to correct for the batch effect, at least to some extent.

```
meta$run <- c(rep(1,9),2,1,1,2,rep(1,10),2,1,1,2)
```

Let us assume that we want to look for significant gene expression differences between tumors treated with DPN and untreated tumors. How should we go about this? Notice that we have four things that vary: patient, time point, treatment, and sequencing run.

Let us prepare the table describing the experiment and the count table:

```
countData <- counts(parathyroidGenes)
```

Filter out lowly expressed genes as explained earlier, this time requiring that the total number of counts is at least the same as the number of samples (so on average, each sample must have at least one count).

```
countData <- countData[rowSums (countData) >= ncol
(countData),]
```

Create the DESeqDataSet object using patient, run, time, and treatment as covariates:

```
dds <- DESeqDataSetFromMatrix(countData = countData,
                colData = meta,
                design = ~patient + run + time + treatment)
```

You can now access the count table in the dds object with counts(dds) and the design formula with design(dds).

Estimate size factors for normalization before running the DESeq command (this step may not be required in all versions of DESeq):

```
dds <- estimateSizeFactors(dds)
```

Run the DESeq command as before. This will build a GLM, fit models and perform tests for various comparisons. The default comparison if you do not specify anything in the results() function will be for treatment differences (because treatment is listed as the last factor in the design formula):

```
dds <- DESeq(dds)
```

To find out which comparisons that have results available, use resultsNames():

```
resultsNames(dds)
```

To obtain results and select the significantly differentially expressed genes for the DPN versus control comparison:

```
res <- results(dds, contrast=c("treatment","DPN",
"Control")) # In older versions of DESeq2, you may
instead need to write:
res<- results(dds,"treatment_DPN_vs_Control")
sig.deseq<- res[which(res$padj<0.01),]
sig.deseq.names<- rownames(sig.deseq)
```

If you had wanted an overall comparison of the 24 and 48 h time points (including all treatments and patients), you could have used `results(dds,"time_48h_vs_24 h")` instead, and so on.

What if we wanted to a more specific or complex comparison? For example, we might want to compare the difference between DPN and control, on the one hand, and the difference between OHT and control, on the other hand, that is, a difference of differences which would tell us something about DPN treatment effects that are not seen in OHT treatment. To do this, we could encode the (DPN vs. control) – (OHT vs. control) using the `contrast` argument of the results function (note that this is only available in DESeq2 versions 1.1.24 and above). As the vector returned by `resultsNames()` has "treatment_DPN_vs_Control" as its seventh entry and "treatment_OHT_vs_Control" as its eighth entry, we can do

```
res<- results(dds,contrast=c(0,0,0,0,0,0,1,-1))
```

8.5.12 For Reference: edgeR Code Example

```
library(edgeR)
d<- DGEList(counts=counts(parathyroidGenes))
d<- d[rowSums(d$counts)>=ncol(counts(parathyroidGe
  nes)),]
d<- calcNormFactors(d,method="TMM")

# For checking if the normalization worked.This plots
the "count-per-million"distributions for each sample
boxplot(cpm(d,normalized.lib.sizes=T),outline=F,las=2)

meta<- pData(parathyroidGenes)
# Replace "run" by original run(1)or rerun (2)
meta$run<- c(rep(1,9),2,1,1,2,rep(1,10),2,1,1,1)
```

```
design <- model.matrix(~treatment+patient+run+time,
  data=meta)

d <- estimateGLMCommonDisp(d,design)
d <- estimateGLMTrendedDisp(d,design)
d <- estimateGLMTagwiseDisp(d,design)
fit <- glmFit(d,design)
lrt <- glmLRT(fit,coef=2)
# The value of coef above depends on the comparison
you are interested in.You can check available
comparisons with colnames(coef(fit)).
temp <- topTags(lrt,n=100000)$table
sig.edger <- temp[temp$FDR < 0.01,]
sig.edger.names <- rownames(sig.edger)

# To compare(DPN vs control)vs(OHT vs control)instead:
lrt <- glmLRT(fit,contrast=c(0,1,-1,0,0,0,0,0)) #
Corresponding to columns 2 and 3 in fit$design
```

8.5.13 Limma Code Example

```
library(limma)
countData <- counts(parathyroidGenes)
meta <- pData(parathyroidGenes)
# Replace "run" byoriginal run(1)or rerun(2)
meta$run <- c(rep(1,9),2,1,1,2,rep(1,10),2,1,1,1)
nf = calcNormFactors(countData,method="TMM")
voom.data <- voom(countData,design=model.matrix(~patie
  nt+run+time+treatment,data=meta),lib.
  size=colSums(countData)*nf)
voom.data$genes<- rownames(countData)
design = model.matrix(~patient+run+treatment+time,
  data=meta)
voom.fitlimma <- lmFit(voom.data,design)
voom.fitbayes <- ebayes(voom.fitlimma)
voom.pvalues <- voom.fitbayes$p.value[,"treatmentDPN"]
voom.adjpvalues <- p.adjust(voom.pvalues,method="BH")
sig.limma.names <- names(which(voom.adjpvalues<0.01))
# To compare (DPN vs control) vs (OHT vs control)
instead:
contrast.matrix <- makeContrasts(treatmentDPN-
  treatmentOHT,levels=design)
voom.fitlimma2 <- contrasts.fit(voom.fitlimma,
  contrast.matrix)
```

```
voom.fitbayes <- ebayes(voom.fitlimma2)
voom.pvalues <- voom.fitbayes$p.value
voom.adjpvalues <- p.adjust(voom.pvalues,method="BH")
sig.limma.names <- rownames(voom.pvalues)[(which(voom.
  adjpvalues<0.01))]
```

ANALYZING DIFFERENTIAL EXPRESSION IN CHIPSTER

You can perform DE analysis in Chipster with Cuffdiff, DESeq(2), and edgeR. Cuffdiff takes BAM files as input, and you can also optionally give it your own GTF file. Here we concentrate on the DESeq and edgeR tools which need a table of counts as input:

- Count reads per genes or transcripts using HTSeq or eXpress, respectively, and combine the count files for all samples to a count table using the tool "Utilities/Define NGS experiment" as described in Chapter 6. In addition to a count table, this generates a phenodata file, which allows you to describe the experimental setup. Using the phenodata editor, mark all samples which belong to the same group with an identical number in the group column. If you have a more complex experimental design, add a new column for each factor.
- Select the count table and one of the DESeq or edgeR tools. Set the desired significance threshold and click "Run." The list of differentially expressed genes is given both as a table (which you can sort by clicking on the header) and as a BED file. The latter allows you to navigate in Chipster genome browser as described in Chapter 4. The result files include also a dispersion plot and an MDS plot which shows the relative similarities between samples.
- If you are using the tool "Differential expression with edgeR for multivariate experiments," please consult the manual pages in order to set the parameters correctly. Note that the result table contains all the genes, and you can filter it based on any column you like using the tool "Filtering/Filter table a column value."

8.6 SUMMARY

Differential expression analysis methods for RNA-seq are still under active development and a set of "best practices" has not quite settled down yet. There are active debates about the best ways to normalize, the virtues of gene-level analysis versus isoform-level analysis, and even what measurement unit to use for reporting gene expression levels.

So which DE analysis method to choose? The isoform-oriented methods (such as BitSeq, Cuffdiff, and ebSeq) are, of course, the appropriate choice if you are interested in the differentially expressed isoforms, and should in theory work better on the gene level also (because the gene-level changes are intimately connected with the isoform-level changes). However, benchmark studies have shown that the popular R/BioConductor packages such as DESeq, edgeR, and limma, using as input data simple counts reads mapping uniquely to genes, tend to perform at least as well. These particular methods also have the added advantage that they can account for more than one varying experimental factor ("covariate") using a generalized linear modeling framework. In experiments with many biological replicates per group, it may also be worth considering nonparametric methods such as SAMSeq and NOISeq. These methods avoid the tricky issue of modeling the read count distribution, but the tradeoff is that they need more replicates to achieve statistical power.

In terms of ease of use, the R/BioConductor packages are all similar, requiring a basic level of proficiency in the R language. The PDF reference manuals and tutorials or "vignettes" for DESeq, limma, and edgeR contain a wealth of information to guide the novice. It should be said that setting up a multifactor analysis takes some practice, but at the end it is worth the effort. Cuffdiff is a Linux/Mac OS X executable and thus requires basic familiarity with command-line work. The actual running of the analysis is a simple one-step process, although running an analysis may take hours or even days and with potentially high demands on memory (RAM). The R-based methods are typically faster, with limma perhaps being the fastest, and less memory-intensive, so that they can be run on an average laptop computer.

REFERENCES

1. Ringnér M. What is principal component analysis? *Nat Biotechnol* 26:303–304, 2008.
2. Müller F.-J., Schuldt B.M., Williams R., Mason D., Altun G., Papapetrou E., Danner S. et al. A bioinformatics assay for pluripotency in human cells. *Nat Methods* 8:315–317, 2011.
3. Auer P.L. and Doerge R.W. Statistical design and analysis of RNA sequencing data. *Genetics* 185:405–416, 2010.
4. Bengtsson M., Ståhlberg A., Rorsman P., and Kubista M. Gene expression profiling in single cells from the pancreatic islets of Langerhans reveals lognormal distribution of mRNA levels. *Genome Res* 15(10):1388–1392, 2005.
5. Anders S. and Huber W. Differential expression analysis for sequence count data. *Genome Biology* 11:R106, 2010.

6. Robinson M.D., McCarthy D.J., and Smyth G.K. edgeR: A Bioconductor package for differential expression analysis of digital gene expression data. *Bioinformatics* 26(1):139–140, 2010.
7. Esnaola M., Puig P., Gonzalez D., Castelo R., and Gonzalez J.R. A flexible count data model to fit the wide diversity of expression profiles arising from extensively replicated RNA-seq experiments. *BMC Bioinformatics* 14(1):254, 2013.
8. Law C.W., Chen Y., Shi W., Smyth G.K. Voom: Precision weights unlock linear model analysis tools for RNA-seq read counts. *Genet Mol* 15: R29, 2014.
9. Love MI, Huber W, and Anders S. Moderated estimation of fold change and dispersion for RNA-Seq data with DESeq2. bioRxiv, doi:10.1101/002832, 2014.
10. Li J. and Tibshirani R. Finding consistent patterns: A nonparametric approach for identifying differential expression in RNA-Seq data. *Stat Methods Med Res* 22(5):519–36, 2013. doi: 10.1177/0962280211428386.
11. Tarazona S., García-Alcalde F., Dopazo J., Ferrer A., and Conesa A. Differential expression in RNA-seq: A matter of depth. *Genome Res* 21(12):2213–2223, 2011. doi: 10.1101/gr.124321.111.
12. Mortazavi A., Williams B.A., McCue K., Schaeffer L., and Wold B. Mapping and quantifying mammalian transcriptomes by RNA-seq. *Nat Methods* 5(7):621–628, 2008. doi: 10.1038/nmeth.1226.
13. Trapnell C., Williams B.A., Pertea G., Mortazavi A., Kwan G., van Baren M.J., Salzberg S.L., Wold B.J., and Pachter L. Transcript assembly and quantification by RNA-seq reveals unannotated transcripts and isoform switching during cell differentiation. *Nat Biotechnol* 28(5):511–515, 2010. doi: 10.1038/nbt.1621.
14. Wang E.T., Sandberg R., Luo S., Khrebtukova I., Zhang L., Mayr C., Kingsmore S.F., Schroth G.P., and Burge C.B. Alternative isoform regulation in human tissue transcriptomes. *Nature* 456(7221):470–476, 2008.
15. Wagner G.P., Kin K., and Lynch V.J. Measurement of mRNA abundance using RNA-seq data: RPKM measure is inconsistent among samples. *Theory Biosci* 131(4):281–285, 2012.
16. Robinson M.D. and Oshlack A. A scaling normalization method for differential expression analysis of RNA-seq data. *Genome Biol* 11(3):R25, 2010. doi: 10.1186/gb-2010-11-3-r25.
17. Dillies M.A., Rau A., Aubert J., Hennequet-Antier C., Jeanmougin M., Servant N., Keime C. et al. on behalf of the FrenchStatOmique Consortium. A comprehensive evaluation of normalization methods for Illumina high-throughput RNA sequencing data analysis. *Brief Bioinform* 14(6):671–683, 2013.
18. Trapnell C., Hendrickson D.G., Sauvageau M., Goff L., Rinn J.L., and Pachter L. Differential analysis of gene regulation at transcript resolution with RNA-seq. *Nat Biotechnol* 31(1):46–53, 2013. doi: 10.1038/nbt.2450.
19. Soneson C. and Delorenzi M. A comparison of methods for differential expression analysis of RNA-seq data. *BMC Bioinformatics* 14:91, 2013.

20. Trapnell C., Roberts A., Goff L., Pertea G., Kim D., Kelley D.R., Pimentel H., Salzberg S.L., Rinn J.L., and Pachter L. Differential gene and transcript expression analysis of RNA-seq experiments with TopHat and Cufflinks. *Nat Protoc* 7(3):562–578, 2012. doi: 10.1038/nprot.2012.016.
21. Glaus P., Honkela A., and Rattray M. Identifying differentially expressed transcripts from RNA-seq data with biological variation. *Bioinformatics* 28(13):1721–1728, 2012.
22. Leng N., Dawson J.A., Thomson J.A., Ruotti V., Rissman A.I., Smits B.M.G., Haag J.D., Gould M.N., Stewart R.M., and Kendziorski C. EBSeq: An empirical Bayes hierarchical model for inference in RNA-seq experiments. *Bioinformatics* 29(8):1035–1043, 2013.
23. Limma user guide. http://www.bioconductor.org/packages/2.12/bioc/vignettes/limma/inst/doc/usersguide.pdf (Accessed 24 October 2013).
24. Bourgon R., Gentleman R., and Huber W. Independent filtering increases detection power for high-throughput experiments. *Proc Natl Acad Sci USA* 107(21):9546–9551, 2010. doi: 10.1073/pnas.0914005107.

CHAPTER 9

Analysis of Differential Exon Usage

9.1 INTRODUCTION

In humans and many other eukaryotes, one gene can be expressed in different forms. Possibly the two most common mechanisms giving rise to these isoforms are alternative promoter usage and alternative splicing. Alternative transcription start sites lead to differences in the beginning of mRNA, whereas alternative splicing causes some of the exons to be skipped and not translated at all. RNA-seq offers exciting possibilities for studying the expression and regulation of isoforms on the whole genome level.

Most of the current RNA-seq methods produce short reads which do not cover full transcripts. Instead, transcripts need to be assembled from sequenced fragments. The assembly and the subsequent abundance estimation can be challenging, because isoforms typically have common or overlapping exons. Furthermore, the coverage along transcripts is not uniform because of biases introduced in sequencing and library preparation. In order to avoid uncertainties in the assembly, one approach for studying alternative isoform regulation of it is to look at differences in the usage of individual exons. The previous chapter discussed differential expression analysis at the gene level, but RNA-seq reads can also be mapped to exons so that the differences in exon-specific counts can be compared between certain conditions, groups, or treatments. This is the topic of this chapter.

The main emphasis of the chapter is on the package DEXSeq from the Bioconductor project [1,2]. The methodology implemented in the DEXSeq package is similar to the methods implemented in the DESeq package that is meant for identification of differentially expressed genes. In addition, certain features of the DEXSeq package are borrowed from the edgeR package. Therefore, the specifics of the methods are not thoroughly discussed here again, but the reader is advised to consult Chapter 8.

Because our aim is to compare some experimental conditions, it is very important to ensure that a suitable number of replicates are generated for each condition. There are some methods that allow for comparison of two different conditions with a single replicate in both groups. One such possibility is a simple comparison of read counts in one gene or transcript at a time using Fisher's exact test. However, this would not allow us to take the biological variation into account, even if we had replicate samples for each group. This approach is implemented in a software MISO [3] which we will not, however, discuss further.

If replicate samples for each group are available, there are more statistically rigorous possibilities for finding the differentially expressed exons between the experimental conditions. The counts of reads could be assumed to follow Poisson distribution, which is a statistical distribution often used for describing count data. In practice, gene- or transcript-specific counts are often overdispersed, and a negative binomial distribution, which can be thought of as a generalization of the Poisson distribution, is a better fit to the data. Thus, the data can be analyzed by a standard generalized linear (regression) model, which takes advantage of the replicate samples to estimate the variance or dispersion for all exons. However, the total number of samples in an RNA-seq experiment is typically small, and the estimates of exon-specific dispersion values would be imprecise. Therefore, the methods implemented in the DEXSeq package make it possible to estimate the dispersion for each exon in such a manner that borrows information from the other similarly expressed exons.

In DEXSeq, a separate negative binomial model is fitted to each gene. The log expression of a gene (or the log of count of reads mapped to it) is modeled as a function of (1) baseline expression of the gene, (2) expected fraction of reads mapped to the gene that are also mapped to a certain exon, (3) log of fold change in a certain condition, and (4) the effect of the condition on the fraction of counts that map to the certain exon. The model lets us identify both the differentially expressed genes and the differentially expressed exons.

A typical analysis workflow with DEXSeq consists of counting reads per exon and reading the count table in R, normalization by estimating size factors, estimation of exon-specific dispersion values, testing for differential exon usage, and visualizing results. These steps are described in detail below using the ENCODE data set in the examples.

9.2 PREPARING THE INPUT FILES FOR DEXSeq

Input files for DEXSeq are generated using two Python scripts that come with the DEXSeq package. In addition, the Python package HTSeq is required. Initially, a flattened GTF file is generated using the Python script dexseq_prepare_annotation.py. Then the counts per each exon (actually per exonic region or exonic bin; see Ref. [2]) are generated using the Python script dexseq_count.py. For a detailed description of the count table generation, see Chapter 6. It is also worth checking the vignette of DEXSeq (a help file giving detailed examples of the usage of the package; available, e.g., from the Bioconductor site), because the details of the process might change. Alternatively, the process can also be done solely using the Bioconductor functionality, as described in Chapter 7 (function `generateCountTable()`), and also in the vignette for the package parathyroidSE.

The Python script dexseq_count.py requires the flattened GTF file and alignments in SAM or BAM formatted file(s). Files containing paired-end data need to be sorted by read name or genomic location. If your files are in BAM format, you can use SAMtools (http://samtools.sourceforge.net/) to convert them into SAM format. A BAM file from the ENCODE project could be converted using the following command:

```
samtools view -h Gm12892_1_chr18.bam -o Gm12892_1_chr18.sam
```

where GM12892_1_chr18.bam is the name of the input file, and the GM12892_1_chr18.sam is the name of the output file.

Each of the SAM files produced above is processed as follows. Note that you also need to indicate whether your data were produced with strand-specific protocol or not (`-s no`).

```
python dexseq_count.py -p yes -s no -r name
Homo_sapiens.GRCh37.70.chr18.chr.gtf Gm12892_1_chr18.sam gm1.txt
```

These commands create seven text files, gm1, …, gm3 and h1, …, h4. Each text file contains two columns, the first one (id) consisting of the Ensembl gene identifier separated from the exon number with a colon, and the second (count) containing the count of reads mapped to that exon (the table does not really contain the header row, it is used here for illustration purposes only):

Id	**Count**
ENSG00000235552:002	35,769
ENSG00000175886:001	15,732
ENSG00000235552:003	7515
ENSG00000235297:001	7275
ENSG00000215492:008	5882

The R functions in the package DEXSeq are seamlessly integrated with the functionality of the HTSeq Python package, and these resulting text files can be directly imported to R, which is what we will cover next.

9.3 READING DATA IN TO R

Reading all the sample-specific count tables is done by the function read. `HTSeqCounts()` from the package DEXSeq. The function expects to get a list of file names that are imported, a name of the flattened GTF file similar to the one used during the generation of the count files, and a data frame describing the experimental setup and the variables that are needed during the statistical analysis of differential exon usage. If the GTF file is not used during the data import, certain visualizations cannot be produced, but statistical testing can still be performed.

Let us first produce a description of the experimental setup. There are three GM samples and four hESC samples in the ENCODE data set. We will first import the GM samples and then the hESC samples. The vector sample name will contain the original sample names, and the vector condition the grouping of the samples to either GM or hESC samples. These vectors are generated first, and then they are bound together to a data frame called phenodata as follows:

```
samplename<-c("Gm12892_1", "Gm12892_2", "Gm12892_3",
              "hESC_1", "hESC_2", "hESC_3", "hESC_4")
condition<-c("gm", "gm", "gm", "esc", "esc", "esc", "esc")
phenodata<-data.frame(samplename, condition)
rownames(phenodata)<-c("gm1", "gm2", "gm3", "h1", "h2", "h3", "h4")
```

Once the phenodata is in place, the count tables can be read into R. Here we will read the data into an ExonCountSet object called ec:

```
library(DEXSeq)
ec<read.HTSeqCounts(countfiles=c("gm1.txt","gm2.txt","gm3.txt",
                                  "h1.txt","h2.txt","h3.txt",
                                  "h4.txt"),
                    design=phenodata)
```

Sometimes the whole exon count data set is in a single table. In such situations, the table can be (1) split up into single files and read into R as above or (2) read into R as a table and converted into an ExonCountSet with the command `newExonCountSet()`. This approach is rather straightforward, and the only hurdle is to generate gene IDs and exon IDs for each row of the table. However, these can be produced from the rownames of the table by splitting them at the colon. The whole process is exemplified below:

```
dat<-read.table("ENCODE_ngs-data-table_exons.tsv", header=T,
                sep="\t")
nc<-nchar(rownames(dat))
geneids<-substr(rownames(dat), 1, nc-4)
exonids<-substr(rownames(dat),nc-2, nc)
ec2<-newExonCountSet(dat, phenodata, geneids, exonids)
```

9.4 ACCESSING THE ExonCountSet OBJECT

After creating the R object containing the data, it is a good practice to check that the object is correctly created and contains the right data in a right format. The phenodata consisting of the sample annotations can be accessed with the function `design()`:

```
design(ec)
```

```
          samplename condition
gm1.txt    Gm12892_1        gm
gm2.txt    Gm12892_2        gm
gm3.txt    Gm12892_3        gm
h1.txt        hESC_1       esc
h2.txt        hESC_2       esc
h3.txt        hESC_3       esc
h4.txt        hESC_4       esc
```

The counts of the reads can be accessed with the function `counts()`. Let us limit the number of rows that are printed to the screen to 10 with the function `head()`:

```
head(counts(ec), 10)
                     gm1.txt  gm2.txt  gm3.txt h1.txt  h2.txt h3.txt h4.txt
ENSG00000000003:E001       0        0        0      0       0      0      0
ENSG00000000003:E002       0        0        0      0       0      0      0
ENSG00000000003:E003       0        0        0      0       0      0      0
ENSG00000000003:E004       0        0        0      0       0      0      0
ENSG00000000003:E005       0        0        0      0       0      0      0
ENSG00000000003:E006       0        0        0      0       0      0      0
ENSG00000000003:E007       0        0        0      0       0      0      0
ENSG00000000003:E008       0        0        0      0       0      0      0
ENSG00000000003:E009       0        0        0      0       0      0      0
ENSG00000000003:E010       0        0        0      0       0      0      0
```

The feature data or the annotations for the genes and exons are accessed with the function `fData()`:

```
head(fData(ec), 10)
                                                              dispBefore
                                geneID  exonID  testable        Sharing  dispFitted
ENSG00000000003:E001 ENSG00000000003    E001        NA             NA          NA
ENSG00000000003:E002 ENSG00000000003    E002        NA             NA          NA
ENSG00000000003:E003 ENSG00000000003    E003        NA             NA          NA
ENSG00000000003:E004 ENSG00000000003    E004        NA             NA          NA
ENSG00000000003:E005 ENSG00000000003    E005        NA             NA          NA
ENSG00000000003:E006 ENSG00000000003    E006        NA             NA          NA
ENSG00000000003:E007 ENSG00000000003    E007        NA             NA          NA
ENSG00000000003:E008 ENSG00000000003    E008        NA             NA          NA
ENSG00000000003:E009 ENSG00000000003    E009        NA             NA          NA
ENSG00000000003:E010 ENSG00000000003    E010        NA             NA          NA

                      dispersion pvalue padjust chr  start end strand transcripts
ENSG00000000003:E001          NA     NA      NA <NA>    NA  NA   <NA>        <NA>
ENSG00000000003:E002          NA     NA      NA <NA>    NA  NA   <NA>        <NA>
ENSG00000000003:E003          NA     NA      NA <NA>    NA  NA   <NA>        <NA>
ENSG00000000003:E004          NA     NA      NA <NA>    NA  NA   <NA>        <NA>
ENSG00000000003:E005          NA     NA      NA <NA>    NA  NA   <NA>        <NA>
ENSG00000000003:E006          NA     NA      NA <NA>    NA  NA   <NA>        <NA>
ENSG00000000003:E007          NA     NA      NA <NA>    NA  NA   <NA>        <NA>
ENSG00000000003:E008          NA     NA      NA <NA>    NA  NA   <NA>        <NA>
ENSG00000000003:E009          NA     NA      NA <NA>    NA  NA   <NA>        <NA>
ENSG00000000003:E010          NA     NA      NA <NA>    NA  NA   <NA>        <NA>
```

Gene Ids are accessed with the function `geneIDs()`. It might be interesting to see how many exons each gene has. Let us count this for a few first genes:

```
data.frame(head(table(geneIDs(ec))))
ENSG00000000003                         15
ENSG00000000005                          9
```

```
ENSG00000000419                          19
ENSG00000000457                          21
ENSG00000000460                          48
ENSG00000000938                          29
```

Function `data.frame()` was used here only in order to get a nicer looking table for the print out. For example, gene ENSG00000000003 has 15 exons.

Similarly, the number of genes having a specific amount of exons can be counted:

```
head(data.frame(table(table(geneIDs(ec)))))
  Var1   Freq
1    1  20436
2    2   7868
3    3   3394
4    4   2068
5    5   1771
6    6   1240
```

There seem to be 20,436 genes with only a single exon in this data set. The maximum number of exons for a “gene” in this data set is 394 exons.

9.5 NORMALIZATION AND ESTIMATION OF THE VARIANCE

The sequencing depth usually differs between the samples, and this bias in the coverage should be taken into account during the analysis. This is accomplished with normalization. The DEXSeq package uses the same normalization methodology as the DESeq package. It tries to normalize the library size and the transcriptome composition between samples. This specific normalization estimates a size factor for each sample in the experiment. Size factors reflect the relative sequencing depths of the different samples.

Size factors are estimated using the function `estimateSizeFactors()`:

```
ec<-estimateSizeFactors(ec)
```

To check what the size factors are, you can use an accessor function `sizeFactors()`:

```
sizeFactors(ec)
  gm1.txt   gm2.txt   gm3.txt    h1.txt    h2.txt    h3.txt    h4.txt
1.6255635 0.7823607 0.7864354 1.1872495 1.0721540 1.0778185 0.8696311
```

After estimating the size factors, we still need to estimate the exon-specific dispersion (variance) before we can perform the actual statistical testing for differential exon usage. The dispersion is estimated from the biological replicates in the data set, but there can also be technical replicates present in the data set. Dispersion cannot usually be estimated for each exon separately, because the number of biological replicates in the data set is typically rather low. Therefore, the estimation of the dispersion is done by borrowing information from the exons that are expressed about at the same rate as the exon the dispersion is being estimated for. This method is often called as dispersion estimation in an intensity-dependent manner.

A dispersion estimate for each gene is calculated with the function `estimateDispersions()`. Calculating the dispersion values could take a fair bit of time, but the progress can be monitored, since a single dot is printed on the screen for each 100 genes that have been processed so far.

```
ec<-estimateDispersions(ec)
Dispersion estimation. (Progress report: one dot per
100 genes)
..
```

In this particular case, the dispersion estimation also gives two warning messages:

```
Warning messages:
1: In .local(object, ...) :
  Exons with less than 10 counts will not be tested.
For more details please see the manual page of
'estimateDispersions', parameter 'minCount'
2: In .local(object, ...) :
  Genes with more than 70 testable exons will be
omitted from the analysis. For more details please see
the manual page of 'estimateDispersions', parameter
'maxExon'.
```

Neither of the warning messages is fatal, and the estimation has been performed successfully. The messages here are to remind you that there are two additional parameters to the function `estimateDispersions()` that limit the number of processed exons. Parameter `minCount` has a default of 10, and only exons with at least the minimum number of counts are processed at this step. In addition, the parameter `maxExons` limits the maximum number of exons in a gene to be no more than 70 by default. If there are more exons in a gene, no dispersion is estimated for it.

The result is stored in the object `ec`'s slot `featureData` column `DispBeforeSharing`. You can access the values with, for example, `featureData(ec)@data$dispBeforeSharing` or `fData(ec)$dispBeforeSharing`.

The dispersion estimates that were just estimated need still to be adjusted so that they borrow information from similarly behaving exons. This is done with the following command that fits a simple function to the dispersion data:

```
ec<-fitDispersionFunction(ec)
```

This also saves the adjusted dispersion estimates into the dispersion column of the featureData slot. After adjusting the dispersion estimated, it is a good practice to check the results. Checking is best done using a plot where the mean of normalized counts is put on the horizontal axis, and the dispersion is on the vertical axis. In addition, a mean dispersion function is added as a line to the plot (Figure 9.1). The plot can be produced with the function `plotDispEsts()`, which is available in

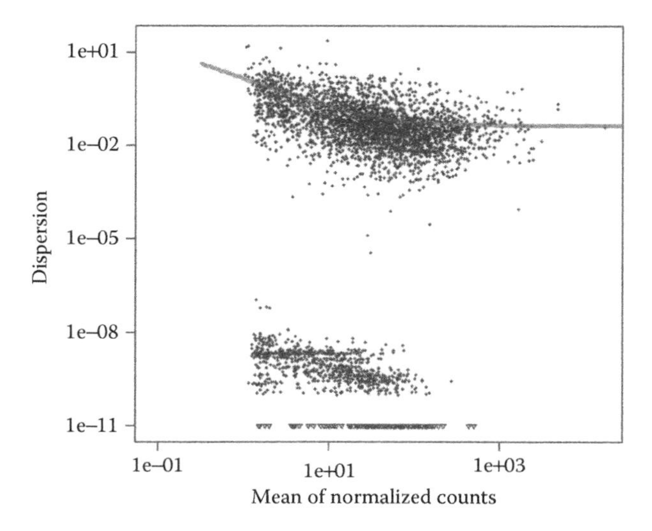

FIGURE 9.1 The mean dispersion plot. Each point in the figure represents one exon. Check that the plotted line (mean dispersion function) follows the shape of the data cloud.

the vignette code of the DEXSeq package and also on the book's code supplement:

```
plotDispEsts(ec)
```

The group of points close to the bottom of the plot are the exons that have a very small estimate of the dispersion. Some of the points form a line to the bottom of the plot, and these have a dispersion of practically zero or exactly zero.

9.6 TEST FOR DIFFERENTIAL EXON USAGE

Once the size factors (normalization) and the dispersion values have been estimated, we have all the components in place for statistical testing of differential exon usage. The testing is based on comparing two generalized linear models with a likelihood test. In the case of the ENCODE data set, there are only two groups that are compared, and the test is rather simple in practice.

For each exon, two different models are fitted. The null model tries to model the expression of an exon from the main effects of sample, exon, and the condition, that is, the group that the sample belongs to. The alternative model takes additionally into account the main effect of a specific exon and its interaction with the condition. Formally, these models can be written as

null model: count ~ sample + exon + condition

alt. model: count ~ sample + exon + condition + exon

= exonID + condition*exon = exonID

A single gene's exons can be tested with the command `testGeneForDEU_BM()`. The command takes two arguments, the exonCountSet object and a gene's name. For example:

```
testGeneForDEU_BM(ec, "ENSG00000017797")
          deviance     df           pvalue
E001     1.4925835      1     2.218161e-01
E002     0.0998600      1     7.519977e-01
E003     1.3569223      1     2.440716e-01
E004     2.6444047      1     1.039151e-01
E005    12.0979356      1     5.047768e-04
E008    26.6814441      1     2.399145e-07
```

```
E009       4.7567963     1       2.918282e-02
E010       8.6114130     1       3.340630e-03
E011       0.8170818     1       3.660348e-01
E012       0.4034728     1       5.253012e-01
E013       0.7998242     1       3.711459e-01
E014       3.0656931     1       7.996105e-02
E015       4.2166483     1       4.002916e-02
E016       4.9595151     1       2.594748e-02
E017       0.1039061     1       7.471916e-01
E018       2.2409165     1       1.344013e-01
E019       0.4172684     1       5.183032e-01
```

The command gives out a table with all the gene's exons on the rows, and the results listed on the columns. The deviance column measures the difference of the goodness of the fit between the null and the alternative models. The deviance values follow a χ^2 distribution, and each exon's p-value is calculated by comparing the deviance to the χ^2 distribution, with df being the number of degrees of freedom. The last column contains the p-values for the exons. Judging from the p-values (≤0.05), four exons (5, 8, 9, and 10) of the gene ENSG00000017797 are differentially expressed between human embryonic stem cells (ecs) and the control cells (gm).

Usually, we are not interested in a single gene and its exons, but we want to test all the genes and exons that are available in the data set. Function `testForDEU()` facilitates this by calling the function `testGeneForDEU_BM()` repeatedly for all the genes in the data set. However, only the genes that were marked testable during the dispersion estimation are tested. The genes that are marked as testable can be found from the feature data table's column testable:

```
head(fData(ec))
                                                    test  dispBefore  disp
                               geneID    exonID  able      Sharing Fitted dispersion
ENSG00000000003:E001 ENSG00000000003    E001 FALSE            NA     Inf      1e+08
ENSG00000000003:E002 ENSG00000000003    E002 FALSE            NA     Inf      1e+08
ENSG00000000003:E003 ENSG00000000003    E003 FALSE            NA     Inf      1e+08
ENSG00000000003:E004 ENSG00000000003    E004 FALSE            NA     Inf      1e+08
ENSG00000000003:E005 ENSG00000000003    E005 FALSE            NA     Inf      1e+08
ENSG00000000003:E006 ENSG00000000003    E006 FALSE            NA     Inf      1e+08

                       pvalue   padjust   chr   start   end  strand  transcripts
ENSG00000000003:E001       NA        NA  <NA>      NA    NA    <NA>         <NA>
ENSG00000000003:E002       NA        NA  <NA>      NA    NA    <NA>         <NA>
ENSG00000000003:E003       NA        NA  <NA>      NA    NA    <NA>         <NA>
ENSG00000000003:E004       NA        NA  <NA>      NA    NA    <NA>         <NA>
ENSG00000000003:E005       NA        NA  <NA>      NA    NA    <NA>         <NA>
ENSG00000000003:E006       NA        NA  <NA>      NA    NA    <NA>         <NA>
```

The actual test is simply done as follows, but it might take a bit of time to complete. If you have just a single chromosome to analyze, you have approximately enough time to fix you cup of coffee. However, if you have a full genome to analyze, now would be a perfect time for a lunch, desert, and coffee.

```
ec<-testForDEU(ec)
Testing for differential exon usage.(Progress report:
one dot per 100 genes)
..
```

In addition to the statistical test result, we can estimate fold changes:

```
ec<-estimatelog2FoldChanges(ec)
```

After we have produced the statistical test results and optionally calculated the fold change values, we can produce a summary table of the combined results:

```
res<-DEUresultTable(ec)
```

Only the statistically significant results can be separated from the not significant results. A first few statistically significant results can be checked as follows:

```
ind<-which(res$padjust <=0.05)
head(res[ind,])

                                geneID  exonID  dispersionp          value
ENSG00000017797:E005 ENSG00000017797    E005    0.14779157  5.047768e-04
ENSG00000017797:E008 ENSG00000017797    E008    0.04967980  2.399145e-07
ENSG00000049759:E007 ENSG00000049759    E007    0.61660215  6.701749e-07
ENSG00000049759:E009 ENSG00000049759    E009    0.07053636  2.911027e-04
ENSG00000049759:E021 ENSG00000049759    E021    0.05289764  2.890047e-03
ENSG00000049759:E038 ENSG00000049759    E038    0.04466578  9.168722e-04

                          padjust   meanBase  log2fold(esc/gm)
ENSG00000017797:E0051.192748e-02   13.30575        -0.8620503
ENSG00000017797:E0081.681800e-05  199.74625         0.4350362
ENSG00000049759:E0074.239748e-05   20.98780        -2.3576794
ENSG00000049759:E0098.002469e-03   50.20427        -0.4834733
ENSG00000049759:E0214.639519e-02  136.85350         0.3015733
ENSG00000049759:E0381.957545e-02  703.55049         0.2016261
```

Some of the genes have a combined gene name, such as ENSG00000119547 + ENSG00000266636. These are the genes that share some exons so that it is not possible to assign the exon to a single gene only. Results for such gene combinations could be difficult to interpret, because the results might stem from differential expression of the genes rather than the differential expression of the exon. For an example of such a result, see the following output:

```
tail(head(res[ind,],69),3)
```

```
                                                                   geneID  exonID
ENSG00000119547+ENSG00000266636:E002 ENSG00000119547+ENSG00000266636    E002
ENSG00000119547+ENSG00000266636:E003 ENSG00000119547+ENSG00000266636    E003
ENSG00000119547+ENSG00000266636:E007 ENSG00000119547+ENSG00000266636    E007
                                      dispersion       pvalue      padjust
ENSG00000119547+ENSG00000266636:E002   0.9580874  6.039439e-06 3.175235e-04
ENSG00000119547+ENSG00000266636:E003   0.1404351  2.284703e-04 6.719902e-03
ENSG00000119547+ENSG00000266636:E007   0.0598415  2.437996e-08 2.050842e-06
                                        meanBase        log2fold(esc/gm)
ENSG00000119547+ENSG00000266636:E002    1.527859               -3.839978
ENSG00000119547+ENSG00000266636:E003   14.307039               -1.286007
ENSG00000119547+ENSG00000266636:E007   81.487062                1.266303
```

Sometimes, it might be beneficial to plot the splicing chart for these specific genes, and see whether the result reflects differential expression of the gene or the exon.

9.7 VISUALIZATION

There are three simple visualizations, MA plot, Volcano plot, and heatmap, that nicely summarize the obtained results. These plots are covered in more detail in Chapter 11. However, there is a specialized MA plot function for the results obtained with DEXSeq package. An MA plot is a normal scatterplot, where the mean expression of an exon is plotted on the x-axis and the fold change on the y-axis.

The MA plot function for an exonCountSet object is available in the supplementary code of this book and in the vignette code for the package DEXSeq. The function marks the exons with an adjusted p-value less than 0.1 with red, and the dots that would fall outside the y-axis limits with triangles. The MA plot can be generated with the code below, and the resulting image is displayed in Figure 9.2.

```
x<-data.frame(baseMean=res$meanBase,
              log2FoldChange=res$'log2fold(esc/gm)',
              padj=res$padjust)
plotMA(na.omit(x), ylim=c(-4,4), cex=0.8)
```

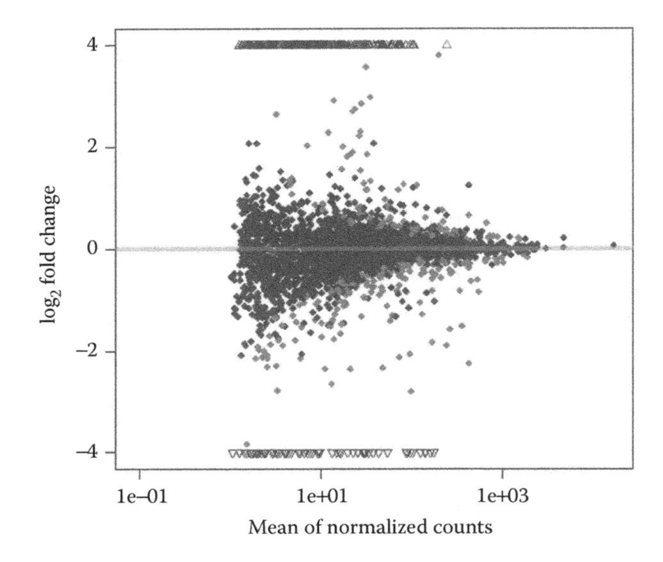

FIGURE 9.2 The MA plot of the results from a differential exon usage analysis.

A plot that can be used for visualizing the exon-specific results for a certain gene is `plotDEXSeq()`. It is possible to plot fitted expression, fitted splicing, or normalized counts in the graph. Unfortunately, the function only works fully if all exon annotations (chromosome, start and end location, strand and transcript names) for all genes in the data set are present in the exonCountSet object. Since not all annotations could be found for the ENCODE data set, the function had to be modified to be able to cope with this. The modified function is present in the chapter-specific extra file on the book website. The modified function is called `plotDEXSeqSimple()` and it works similarly to the `plotDEXSeq()` function, but it only plots fitted expression values, and, if all exons of the gene to be plotted are fully annotated, also the transcript structures.

To generate a plot for a single gene, the function `plotDEXSeqSimple()` expects to get the name of the exonCountSet, the name of the gene to plot as a character vector, and a logical indicator whether to plot a legend or not. The following command generates the result seen in Figure 9.3:

```
plotDEXSeqSimple(ec, "ENSG00000226742", legend=TRUE)
```

An example of a gene that contains a differentially expressed exon is ENSG00000119541, and the results plotted for it with the code below are displayed in Figure 9.4.

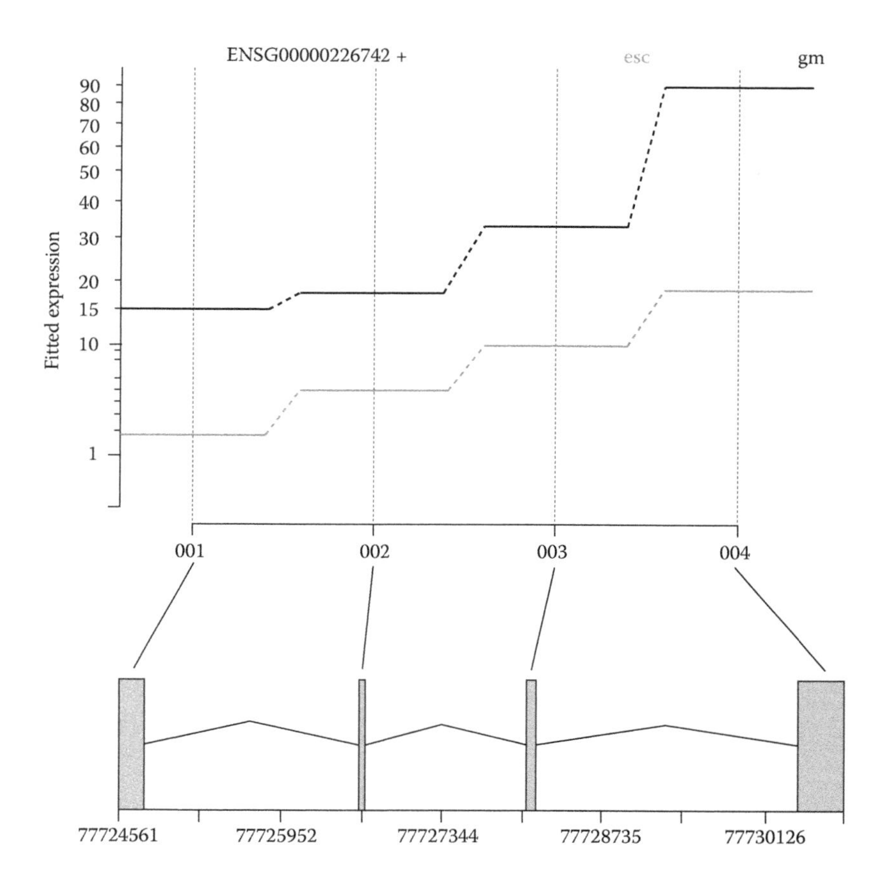

FIGURE 9.3 A plot of result obtained with DEXSeq package. In the upper panel, the expression of the exons of the gene ENSG00000226742 in the two conditions (esc and gm) is shown with two lines. Every exon is clearly more expressed in the gm cells than in the esc cells. Therefore, the result is probably due to differential expression of the whole gene. True enough, the result table verifies this (`res[res$geneID=="ENSG00000226742",]`), since none of the exons are statistically significantly differentially expressed. The lower panel displays the exonic structure of the gene. Exons are displayed as gray bars, and introns as black wedged lines between the exons.

```
plotDEXSeqSimple(ec, "ENSG00000119541", legend=TRUE)
```

In addition to a single plot, plots can be generated to all significant results in HTML format with a simple command:

```
DEXSeqHTML(ec, as.character(unique(res[ind,1])))
```

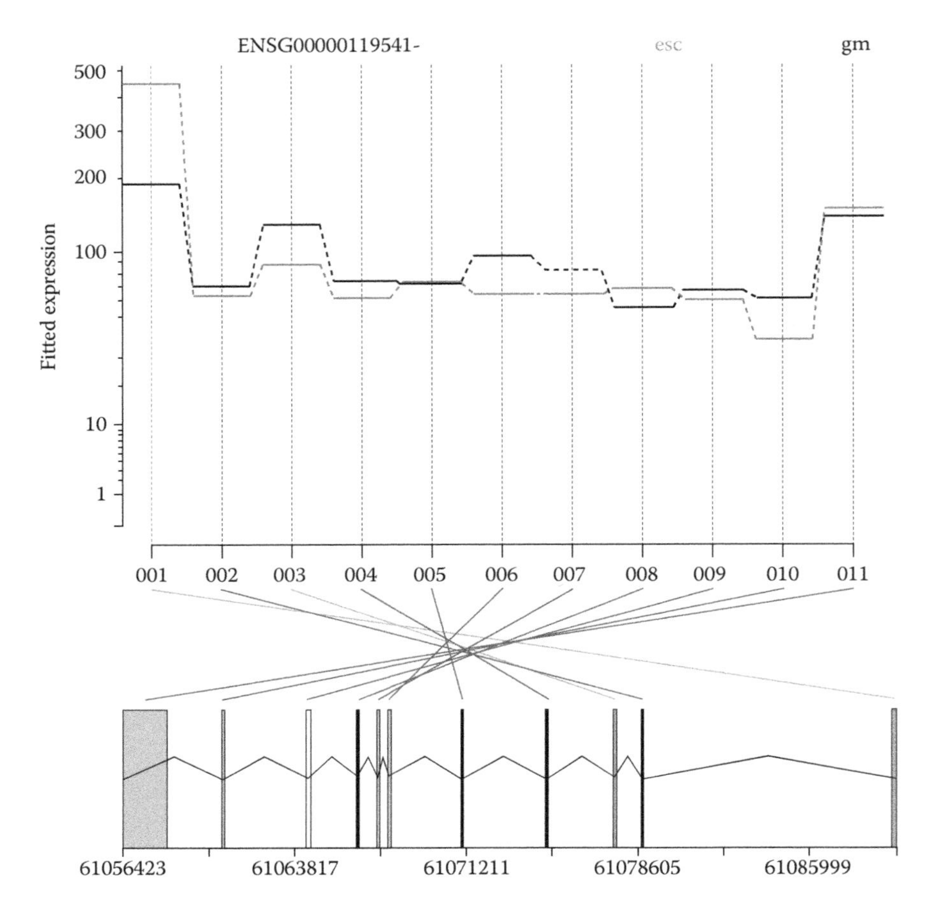

FIGURE 9.4 Results for gene ENSG00000119541. Exon 1 clearly differentially expressed between the esc and the gm cell lines. Another exon is also statistically significantly differentially expressed. Can you guess which one it is? (It is exon number 3.)

The command generates a directory (folder) to the current working directory. The folder contains a single summary webpage and a subfolder with the individual, gene-specific plots. The first argument to the function `DEXSeqHTML()` is the exonCountSet, and the second is a vector of gene names that specifies which genes should be plotted.

The summary page is shown in Figure 9.5.

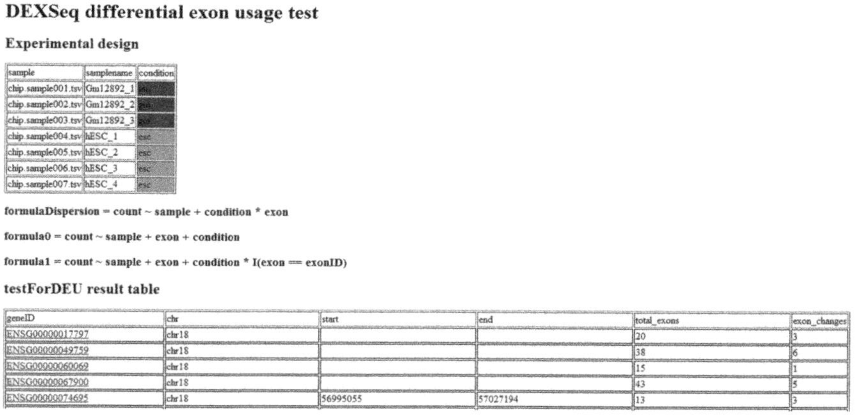

DEXSeq differential exon usage test

Experimental design

sample	samplename	condition
chip.sample001.tsv	Gm12892_1	[illegible]
chip.sample002.tsv	Gm12892_2	[illegible]
chip.sample003.tsv	Gm12892_3	[illegible]
chip.sample004.tsv	hESC_1	esc
chip.sample005.tsv	hESC_2	esc
chip.sample006.tsv	hESC_3	esc
chip.sample007.tsv	hESC_4	esc

formulaDispersion = count ~ sample + condition * exon

formula0 = count ~ sample + exon + condition

formula1 = count ~ sample + exon + condition * I(exon == exonID)

testForDEU result table

geneID	chr	start	end	total_exons	exon_changes
ENSG00000017797	chr18			20	3
ENSG00000049759	chr18			38	6
ENSG00000060069	chr18			15	1
ENSG00000067900	chr18			43	5
ENSG00000074695	chr18	56995055	57027194	13	3

FIGURE 9.5 A summary page of the HTML report generated by the DEXSeq package. Information on the samples, and the exact models fitted to the data are shown. The table in the bottom contains information for individual genes, and the first column of each line of the table in a link to a more detailed result page for that particular gene.

ANALYZING DIFFERENTIAL Exon EXPRESSION IN CHIPSTER

- Count reads using the tool "RNA-seq/Count aligned reads per exons for DEXSeq," and combine the count files for all samples to a count table using the tool "Utilities/Define NGS experiment" as described in Chapter 6. In addition to a count table, this generates a phenodata file, which allows you to describe the experimental setup. Using the phenodata editor, mark all samples which belong to the same group with an identical number in the group column.
- Select the count table and the tool "RNA-seq/Differential exon expression using DEXSeq." In the parameters, set the organism and the desired *p*-value threshold, and click "Run."
- The statistical testing results are reported in two files: one for all the genes, and another for genes which contain differentially expressed exons. The result files also include an MA plot, a dispersion plot, and a visualization of genes which contain differentially expressed exons.

9.8 SUMMARY

A gene can produce multiple transcript isoforms by alternative promoter usage and alternative splicing. RNA-seq offers exciting possibilities for studying the expression and regulation of isoforms on the whole genome level. However, the assembly and abundance estimation of transcript isoforms is complicated by common exons and nonuniform transcript coverage. In order to avoid uncertainties introduced by transcript assembly, one approach for studying alternative isoform regulation is to look at differences in the usage of individual exons.

The DEXSeq package allows the user to test for differential exon usage with a statistical model fitted to the data using a negative binomial distribution. DEXSeq shares some features with the DESeq package and some ideas with the edgeR package. DEXSeq also contains some functionality for plotting the results of the analysis. After the DEXSeq analysis, the results could be further annotated or visualized using the functionality offered by the Bioconductor project.

REFERENCES

1. Anders S. HTSeq documentation website, 2012. http://www-huber.embl.de/users/anders/HTSeq/doc/overview.html (Accessed 17 January 2014).
2. Anders S., Reyes A., and Huber W. Detecting differential usage of exons from RNA-seq data. *Genome Res* 22:2008, 2012. doi:10.1101/gr.133744.111
3. Katz Y., Wang E.T., Airoldi E.M., and Burge C.B. Analysis and design of RNA sequencing experiments for identifying isoform regulation. *Nat Methods* 7:1009–1015, 2010.

CHAPTER 10

Annotating the Results

10.1 INTRODUCTION

In general, an annotation is any comment, explanation, or some markup (collectively known as metadata) that is attached to data. Metadata are often attached to some specific part of the data, but they can also describe how, where, or when the data were collected. However, in bioinformatics, the meaning of annotation is more specific. Genome annotation refers to the process of identifying and locating the genes and other functional elements of an organism's genome and attaching some notes of their functions on them. Typical annotations that are associated with a gene are the genomic location (chromosome, cytoband, base pairs), the exonic and intronic structure, the transcripts, and some functional notions possibly in the form of controlled vocubularies, such as gene ontology (GO), or the metabolic pathways that the translated proteins function in (e.g., KEGG, the Kyoto Encyclopedia of Genes and Genomes; Reactome, a database of reactions, pathways, and biological processes).

The use of annotations in an RNA-seq experiment is typically two-fold. If the mapping of the reads is done against a reference genome, the annotations are used in this mapping phase to assign the reads to the correct genes and/or transcripts. After the data analysis, more detailed annotations are typically retrieved for the interesting or differentially expressed transcripts so that a biological meaning can be postulated for the observed results. For example, using the functional information of the genes and transcripts, such as gene ontology categories, allows for more refined analyses of the up- or downregulated metabolic pathways.

This chapter covers retrieval of additional annotations (reannotation) for the differentially expressed genes and some analyses that these new annotations make possible.

10.2 RETRIEVING ADDITIONAL ANNOTATIONS

There is a plethora of biological databases that contain annotations for genes, transcripts, and their products. Nowadays annotations are often gathered together from individual primary databases, such as GenBank (http://www.ncbi.nlm.nih.gov/genbank/), into secondary databases, possibly most notably genome databases, such as UCSC Genome Browser (http://genome.ucsc.edu/) and Ensembl (http://www.ensembl.org/). BioMart's Bio Portal (http://www.biomart.org/) offers a convenient interface to retrieve information from the Ensembl and over 40 other databases.

It is possible to add annotations from these and many other sources to the differentially expressed genes either by querying each database separately and then assembling a comprehensive data set by hand or by using the readily available Bioconductor packages that query only certain databases, typically those that have the basic information on the transcripts and their protein products.

Each gene or transcript is usually associated with an accession number that points to that specific feature in some database. For example, each gene has a separate and unambiguous accession number in the Ensembl database. Gene-specific accession number in the Ensembl database starts with the string ENSG, and transcript-specific accession numbers start with ENST. In addition to the accession numbers, it is possible to query the databases with the gene or transcript sequence, something that a BLAST search would accomplish. There is a tailor-made service, Blast2GO (http://www.blast2go.com/b2ghome) that tries to find the relevant functional annotation to known or even previously unknown sequences. Sometimes, it is best to use the gene names to find the annotations, but unfortunately that route is seldom very simple. Genes can have several commonly used names, and even if the gene names are somewhat standardized in humans, there are still typically many synonyms for each gene. Cross-species comparisons with the gene names are not typically very reliable due to this ambiguous naming strategy. In summary, it is best to resort to the accession number when searching for additional information, if the accession number is available.

Using accession number for annotation of the gene and transcript information is rather straightforward. Sometimes, accession numbers need to be translated from one database convention to another, which might lead to some attrition and data loss, because not all accession numbers typically have a direct pair in the other database. For example, the estimated number of the human genes is still different in the Ensembl and the UCSC Genome browser. This stems from the slightly different genome annotation processes.

Let us next retrieve some extra information for the genes found to be differentially expressed in Chapter 9. This new information will be accessed using a couple of Bioconductor packages.

10.2.1 Using an Organism-Specific Annotation Package to Retrieve Annotations for Genes

One of the simplest methods to add some new annotations to the genes is to get them from an organism-specific annotation package in R. For humans, such a package is org.Hs.eg.db, and there are also similar packages for many model organisms, all containing a prefix org. Organism-specific packages allow conversion of certain accession numbers to other database's accession numbers, and also contain basic information about the gene's location in the genome, its functional annotations using gene ontology classes, etc. Let us find out the GO classes for the differentially expressed genes identified in Chapter 9.

Human-specific package is based on Entrez Gene accession numbers, but it contains Ensembl IDs also. The differentially expressed genes identified in Chapter 9 contain Ensembl IDs. Therefore, before attaching the GO classes to the genes, we need to convert Ensembl IDs into Entrez Gene IDs, and then using these new accession numbers retrieve the GO classes from the annotation package. Some loss of information might be expected in the process.

The object `res` contains the results of the differential expression analysis performed in Chapter 9. It contains several rows per every gene because the statistical analysis was actually performed for the exons, not genes themselves. Let us first extract the genes that were statistically significantly differentially expressed into an object `res2` and then extract all the unique Ensembl gene identifiers from that result set into an object `deg`:

```
res2<-res[!is.na(res$padjust) & res$padjust<=0.05,]
deg<-as.character(unique(res2$geneID))
```

There are totally 87 differentially expressed genes in the set. The actual number might vary a bit, but with the same version of R and its add-on packages on a Windows 7 system should produce the same number of genes. These can be converted into Entrez Gene identifiers using the environment org.Hs.egENSEMBL2EG from the human-specific package org.Hs.eg.db. It contains a mapping from Ensembl gene IDs to Entrez Gene IDs.

All the available fields in the annotation package can be checked with the command `ls()`:

```
library("org.Hs.eg.db")
ls("package:org.Hs.eg.db")
```

```
 [1] "org.Hs.eg"                 "org.Hs.eg.db"
 [3] "org.Hs.eg_dbconn"          "org.Hs.eg_dbfile"
 [5] "org.Hs.eg_dbInfo"          "org.Hs.eg_dbschema"
 [7] "org.Hs.egACCNUM"           "org.Hs.egACCNUM2EG"
 [9] "org.Hs.egALIAS2EG"         "org.Hs.egCHR"
[11] "org.Hs.egCHRLENGTHS"       "org.Hs.egCHRLOC"
[13] "org.Hs.egCHRLOCEND"        "org.Hs.egENSEMBL"
[15] "org.Hs.egENSEMBL2EG"       "org.Hs.egENSEMBLPROT"
[17] "org.Hs.egENSEMBLPROT2EG"  "org.Hs.egENSEMBLTRANS"
[19] "org.Hs.egENSEMBLTRANS2EG" "org.Hs.egENZYME"
[21] "org.Hs.egENZYME2EG"        "org.Hs.egGENENAME"
[23] "org.Hs.egGO"               "org.Hs.egGO2ALLEGS"
[25] "org.Hs.egGO2EG"            "org.Hs.egMAP"
[27] "org.Hs.egMAP2EG"           "org.Hs.egMAPCOUNTS"
[29] "org.Hs.egOMIM"             "org.Hs.egOMIM2EG"
[31] "org.Hs.egORGANISM"         "org.Hs.egPATH"
[33] "org.Hs.egPATH2EG"          "org.Hs.egPFAM"
[35] "org.Hs.egPMID"             "org.Hs.egPMID2EG"
[37] "org.Hs.egPROSITE"          "org.Hs.egREFSEQ"
[39] "org.Hs.egREFSEQ2EG"        "org.Hs.egSYMBOL"
[41] "org.Hs.egSYMBOL2EG"        "org.Hs.egUCSCKG"
[43] "org.Hs.egUNIGENE"          "org.Hs.egUNIGENE2EG"
[45] "org.Hs.egUNIPROT"
```

There are 45 different fields available in the package, and which one to use naturally depends on the information we are interested in. Each one of these fields has a help page, and we selected the org.Hs.egENSEMBL2EG field, because it contains the largest amount of GO annotation information for the genes.

To continue the workflow, first the mapping information is extracted from the package into an object xx, which is then further converted into a data frame xxd:

```
xx <- as.list(org.Hs.egENSEMBL2EG)
xxd <- as.data.frame(unlist(xx))
```

The contents of the object xxd should look similar to:

```
head(xxd)
                    unlist(xx)
ENSG00000121410              1
ENSG00000175899              2
ENSG00000256069              3
ENSG00000171428              9
ENSG00000156006             10
ENSG00000196136             12
```

The rows that contain the Entrez Gene IDs for the differentially expressed genes can be extracted with a simple subsetting after generating an index for the parts that need to be extracted:

```
ind<-match(deg, rownames(xxd))
degeg<-xxd[ind,]
```

There were originally 87 differentially expressed genes, but an Entrez Gene ID was found only for 74 genes. Next, we can search for the GO classes for these 74 genes. GO class information is contained in the environment org.Hs.eg.dbGO in the package org.Hs.eg.db. Similar to what we did for the identifiers, we first extract the GO classes into a data frame, which we then subset with an index, and the final result is saved into a list degeggo:

```
xx <- as.list(org.Hs.egGO)
ind<-match(degeg, names(xx))
degeggo<-xx[ind]
```

The problem with further using the GO classes for downstream analyses is that a certain gene can map to several GO classes, and consolidating these results into a simple table or a data frame, that would be easy to manipulate, is not always straightforward.

However, there is a simple way of getting the results into a table. The package AnnotationDbi contains the function `select()` that can be used for selecting annotations from the organism-specific annotation packages. It requires four arguments: the name of the annotation package, the accession numbers to be searched for, the fields to be extracted, and the name of the annotation package. The available fields in the package can be found with the command `keytypes()`, for example,

```
library(AnnotationDbi)
keytypes(org.Hs.eg.db)
```

```
 [1] "ENTREZID"    "PFAM"        "IPI"         "PROSITE"      "ACCNUM"
 [6] "ALIAS"       "CHR"         "CHRLOC"      "CHRLOCEND"    "ENZYME"
[11] "MAP"         "PATH"        "PMID"        "REFSEQ"       "SYMBOL"
[16] "UNIGENE"     "ENSEMBL"     "ENSEMBLPROT" "ENSEMBLTRANS" "GENENAME"
[21] "UNIPROT"     "GO"          "EVIDENCE"    "ONTOLOGY"     "GOALL"
[26] "EVIDENCEALL" "ONTOLOGYALL" "OMIM"        "UCSCKG"
```

Every field can be mapped to any of the others, so we can map the differentially expressed genes to GO categories either by using the Entrez Gene IDs or the Ensembl Gene IDs. This can be achieved by either of the following commands:

```
degeggo<-select(org.Hs.eg.db, as.character(degeg),
                "GO", keytype = "ENTREZID")
degeggo<-select(org.Hs.eg.db, as.character(deg),
                "GO", keytype = "ENSEMBL")
```

The output from this command retrieves possibly several rows for each gene, one GO category per line:

```
head(degeggo)
          ENSEMBL         GO  EVIDENCE  ONTOLOGY
1  ENSG00000017797 GO:0005096       IDA        MF
2  ENSG00000017797 GO:0005515       IPI        MF
3  ENSG00000017797 GO:0005829       TAS        CC
4  ENSG00000017797 GO:0006200       IDA        BP
5  ENSG00000017797 GO:0006810       IDA        BP
6  ENSG00000017797 GO:0006935       TAS        BP
```

Each GO class is also associated with an evidence code that is documented at http://www.geneontology.org/GO.evidence.shtml. Of the ones shown above, both IDA and IPI are experimentally validated, and TAS is a tracable author statement. Each class also belongs to one of the three ontology classes, biological process (BP), molecular function (MF), and cellular component (CC), which are listed in the column ONTOLOGY. The data are now in a format that we need for further analyses.

If we want to do some comparison analysis later on to find whether some of the GO categories are enriched in the differentially expressed genes, we might want to generate a universe gene list that contains all the genes from the chromosome 18 (the ENCODE experiment only included data for the chromosome 18) and their GO annotations. The res object (from Chapter 9) contains the differentially expressed genes, so we want to retrieve the GO annotations with the chromosome information for these genes. Then we filter away the genes that are not located into chromosome 18 or have a missing chromosome assignment and remove the chromosome information column from the table:

```
univ<-as.character(unique(res$geneID))
univgo<-select(org.Hs.eg.db, univ, c("CHR", "GO"),
                    keytype="ENSEMBL")
univgo<-univgo[univgo$CHR=="18" & !is.na(univgo$CHR),
                    c(1, 3:5)]
```

10.2.2 Using BioMart to Retrieve Annotations for Genes

The biomaRt package from the Bioconductor project provides an interface to the BioMart database collection. Before any queries, a database need to be selected. A list of all marts can be generated with the command:

```
library(biomaRt)
listMarts()
```

Among the marts you should see Ensembl that we are going to use for retrieving the annotations for the human genes. Next a connection to the database has to be established using the function `useMart()`, which takes as an argument the name of the mart:

```
ensembl=useMart("ensembl")
```

After selecting the database, a data set need to be selected. For Ensembl, these are organisms based. Available data sets can be listed with the function `listDatasets()`:

```
listDatasets(ensembl)
```

For humans, the data set is called hsapiens_gene_ensembl, and its version at the time of writing was GRCh37.p12. Once the data set is selected, the ensemble object can be updated to include this information:

```
ensembl = useDataset("hsapiens_gene_ensembl",
                     mart= ensembl)
```

Once the connection is correctly established, the annotation information need to be selected, and similarly, what filters need to be applied, if any. For example, we are going to download the Ensembl accession number, chromosome, and GO categories for each of the differentially expressed genes. These are collectively called attributes. In addition, we can restrict the search to the chromosome 18, only. The available attributes and filters can be listed with the functions `listAttributes()` and `listFilters()`. Since there are hundreds of both, we will save them in two objects for easier browsing:

```
filters=listFilters(ensembl)
attributes=listAttributes(ensembl)
```

Checking the attribute list, the attributes we need to use are ensemble_gene_id, chromosome_name, and go_id. We also use chromosome_name as a filter, because we only want to get the annotations for the genes in the chromosome 18. The database query is submitted with the function `getBM()`, which takes four arguments: attributes, filters, values, and mart. The attributes and filters were already covered; the values argument should take the list of accession numbers, which are currently in the object deg, and additionally the name of the chromosome that is used as a filter (saved in an object chrom); the mart is the object ensemble created above. Thus, the query in its entirety would be:

```
chrom<-c(18)
query<-getBM(attributes=c("ensembl_gene_id",
                          "chromosome_name",
                          "go_id"),
             filters="chromosome_ name",
             values=list(deg, chrom),
             mart=ensembl)
```

Once the query is ready, and R returns to the prompt, we can check a couple of rows from the beginning of the resulting object query:

```
head(query)
 ensembl_gene_id  chromosome_name       go_id
```

```
1ENSG00000101574                    18        GO:0006139
2ENSG00000101574                    18        GO:0003676
3ENSG00000101574                    18        GO:0008168
4ENSG00000101574                    18
5ENSG00000154065                    18        GO:0005515
6ENSG00000080986                    18        GO:0008608
```

There can be several lines for each gene, because as above in the case of using the organism-specific package, there can be several GO categories assigned to a gene, and each of these are put on their own lines.

There were 87 genes in the original gene list stored as the object deg. If you check the number of unique genes that the query returned, you will see that more genes were returned that were in the original gene list:

```
length(unique(query$ensembl_gene_id))
[1]289
```

This indicates that if we want to get the results for the differentially expressed gene only, we need to add those as a filter in the query, for example,

```
query<-getBM(attributes=c("ensembl_gene_id",
                          "chromosome_name",
                          "go_id"),
             filters=c("ensembl_gene_ id",
                       "chromosome_name"),
             values=list(deg,chrom),
             mart=ensembl)
length(unique(query$ensembl_gene_id))
[1]72
```

We can run the same to get annotations for all genes assigned to the chromosome 18:

```
query2<-getBM(attributes=c("ensembl_gene_id",
                           "chromosome_name",
                           "go_id"),
              filters=c("chromosome_ name"),
              values=list(chrom),
              mart=ensembl)
```

Note that the filters need to be in the same order as the values! Here, the Ensemble gene IDs are first, followed by the chromosome name in both arguments.

10.3 USING ANNOTATIONS FOR ONTOLOGICAL ANALYSIS OF GENE SETS

There are several terms commonly used for methods that test whether genes in an expression profiling experiment are somehow enriched or overrepresented in some biologically meaningful classes, such as GO ontology categories, biological (metabolic) pathways, or other functional groups. Commonly used terms are gene set analysis (GSA), gene enrichment analysis, or ontological analysis. These methods have been reviewed in several publications [1–3].

A 2×2 table is often employed in the tests for overrepresentation. For the whole experiment, genes are first assigned to either overexpressed or not overexpressed groups, which are then both further divided into parts that either are in a certain functional group or are not. Thus, a simple table can be constructed, where the number of genes in each cell of the table is represented by letters a–d:

	Overexpressed	**Not Overexpressed**
In the functional group	a	b
Not in the functional group	c	d

Let us take a concrete example using the annotations we have previously acquired into objects query and query2. There a total of 72 differentially expressed genes, and 289 genes altogether in the chromosome 18. This does not actually reflect the real number of genes in the chromosome 18, but it is the number biomaRt returns. The GO category 0005515 means a protein-associated chaperone, and the following code counts how many genes as associated with that category in both gene lists (query and query2). First we will find the rows of the annotation table that hold genes with the GO annotation to the category 0005515, and save these into separate indicator vectors:

```
indq<-which(query$go_id=="GO:0005515")
indq2<-which(query2$go_id=="GO:0005515")
```

Using the indicators, we can count the number of different genes in both gene lists that have been annotated to that category. The column ensemble_gene_id holds the gene identifiers that are unambiguous. Selecting only the unique gene identifiers will then tell the count we are after:

```
length(unique(query$ensembl_gene_id[indq]))  # 44
length(unique(query2$ensembl_gene_id[indq2])) # 132
```

There are 132 genes in the chromosome 18 associated with that GO category and 44 genes that were differentially expressed. The 2 × 2 table for assessing the overrepresentation then becomes:

	Overexpressed	**Not Overexpressed**
In the functional group	44	132
Not in the functional group	72 – 44 = 28	289 – 132 = 157

We can represent this table in R as a matrix using the function `matrix()`:

```
mat<-matrix(ncol=2, data=c(44,28,132,157))
mat

      [,1]   [,2]
[1,]    44    132
[2,]    28    157
```

In its simplest form, the overrepresentation can be tested with Fisher's exact text, which in R can be performed for a matrix simply with the function `fisher.test()`:

```
fisher.test(mat)

      Fisher's Exact Test for Count Data
data: mat
p-value = 0.02471
alternative hypothesis: true odds ratio is not equal
to 1
95 percent confidence interval:
 1.069865 3.296576
sample estimates:
odds ratio
 1.86583
```

The *p*-value for the test is smaller than the classically used for cut-off statistical significance (0.05). Thus, the GO category protein-associated chaperone would appear to be overrepresented among the differentially expressed genes.

The following example gives a rough overview of how the analysis is done for a single functional category. In practice, the analysis should be repeated for each category separately. Fisher's test covered above is just one of the many different possibilities for the gene set analysis and other methods are covered in slightly more details in the rest of the chapter.

10.4 GENE SET ANALYSIS IN MORE DETAIL

Goeman and Bühlmann (2005) divide the different analysis methods into competitive and self-contained tests. In competitive methods, all the genes in the experiment are first divided into the genes that are in a certain group, such as one GO category, and all the other genes that are not in this group. Then the number of differentially expressed genes in both groups is compared, and if the result is significant, the result indicates that the GO category is activated. Self-contained tests use only the information of the genes in a certain group, such as a GO category, and test a hypothesis that none of the genes in the group are differentially expressed.

Furthermore, Goeman and Bühlmann (2005) divide the tests from a different perspective: the sampling unit. In typical tests that employ the 2×2 table, the gene is used as sampling unit, that is, it is the genes that are divided into the cells of the table. This is in contrast to the usual statistical practice, where samples or individuals are used as sampling units. Goeman and Bühlmann (2005) make this distinction between gene-sampling and subject-sampling models. Self-contained methods are usually based on subject sampling, and competitive methods are typically grounded in gene sampling.

Both Goeman and Bühlmann (2005) and Maciejewski (2013) argue that methods that are self-contained and based on subject sampling are more valid than others and also offer easy interpretablity. Algorithms that use this kind of methods are, for example, SAFE [4] and Globaltest [5], both released as Bioconductor packages.

Competitive methods are available in Bioconductor packages GOstats, topGO, and several others, just to mention a few: Bioconductor package goseq implements a method that corrects for the possible length bias inherent for the RNA-seq data. Package GSVA offers a method that assesses relative enrichment of pathways across the whole experiment. Limma package contains a method that converts the RNA-seq data into a similar scale as the DNA microarray data (voom) and then a couple of gene set enrichment methods, namely roast and camera.

The RNA-seq-specific analysis methods are not that widely available, and at the moment, the best bet might be to treat RNA-seq like one would treat DNA microarray data, naturally after the needed normalization (e.g., limma/voom or edgeR) and transformation (variance stabilizing or log) steps. As an example of possibilities, let us next apply a competitive method, self-contained method, and a length bias correcting method for the RNA-seq data. For the following examples, we are going to use the parathyroid data set, and results based on genes, not exons.

Before the examples of different methods, let us generate a list of differentially expressed genes for the parathyroid data set. The following code block runs an analysis using edgeR package and follows the steps that have been covered in more details in Chapter 8. The results are a table with fold changes and p-values (sig.edger) and a vector of Ensembl IDs for differentially expressed genes (sig.edger.names):

```
# Differential analysis using edgeR
library(edgeR)
library(parathyroid)
data(parathyroidGenes)
d<-DGEList(counts=counts(parathyroidGenes))
d<-d[rowSums(d$counts)>=
      ncol(counts(parathyroidGenes)),]
d<-calcNormFactors(d,method="TMM")
meta <- pData(parathyroidGenes)
design <- model.matrix(~treatment+time, data=meta)
d <- estimateGLMCommonDisp(d,design)
d <- estimateGLMTrendedDisp(d,design)
d <- estimateGLMTagwiseDisp(d,design)
fit <- glmFit(d,design)
# Differential expression through time
lrt <- glmLRT(fit,coef=4)
temp <- topTags(lrt,n=100000)$table
sig.edger <- temp[temp$FDR < 0.01,]
sig.edger.names <- rownames(sig.edger)
```

10.4.1 Competitive Method Using GOstats Package

The Bioconductor package GOstats has been around for quite some time and typically works reliably when a list of differentially expressed genes is available. In addition, a separate list of genes in the universe, such as an organism's genome, is needed. The data preprocessing is done in two steps: First, all the genes in the organism-specific annotation packages, here org.Hs.eg.db, are saved into object reference.genes as Entrez Gene IDs. Then the Ensembl IDs of the differentially expressed genes (in object sig.edger.names) are converted into Entrez Gene IDs using the annotation package org.Hs.eg.db. The code that performs these manipulations is shown below:

```
library(org.Hs.eg.db)
library(GOstats)
ensembl.to.entrez <- as.list(org.Hs.egENSEMBL2EG)
reference.genes <- unique(unlist(ensembl.to.entrez))
selected.genes<-na.omit(
                      unique(
                        select(
                           org.Hs.eg.db,
                           sig.edger.names,
                           c("ENTREZID"),
                           keytype="ENSEMBL")$ENTREZID
                        )
                      )
```

The GOstats package allows one to test for over- or underrepresentation of the GO terms among the differentially expressed genes compared to the universe of genes. Usually an analysis for overrepresentation is performed. Over- or underrepresentation can be tested taking the hierarchical nature of the GO ontology into account, because it often happens that if a higher level is statistically significantly overrepresented, its children are also often overrepresented. This is called conditional testing in GOstats. In addition, the one ontology needs to be selected from biological process (BP), molecular function (MF), and cellular component (CC), and a suitable *p*-value for a cut-off for statistical significance needs to be set. The analysis is initialized by setting the parameters for the analysis, and a list of differentially expressed genes and the gene universe as well as the annotation package from where these are available need to be set. For example, the following code does the setup for our intended analysis:

```
params <- new('GOHyperGParams', geneIds=selected.genes,
                universeGeneIds=reference.genes,
                annotation='org. Hs.eg.db', ontology='BP',
                pvalueCutoff=0.01, conditional=TRUE,
                testDirection="over")
```

The analysis is actualy run with the command `hyperGTest()`, and the results given by it can be processed into a nicer table using the command `summary()`:

```
go <- hyperGTest(params)
go.table <- summary(go, pvalue=2)
```

The object `go.table` now holds the results for all possible GO categories:

```
head(go.table)
```

```
      GOBPID        Pvalue   OddsRatio   ExpCount  Count  Size
1 GO:0048285  3.061584e-10    6.398889  3.8920412     21   373
2 GO:0051783  1.386607e-05    8.197938  1.0956148      8   105
3 GO:0006950  1.654383e-05    2.113116 32.3467230     55  3100
4 GO:0000070  1.792116e-05   12.560139  0.5530246      6    53
5 GO:0007067  2.069431e-05    4.829758  2.7525349     12   279
6 GO:0010564  2.560388e-05    6.435160  1.5518320      9   155
                                     Term
1                        organelle fission
2           regulation of nuclear division
3                       response to stress
4    mitotic sister chromatid segregation
5                                  mitosis
6          regulation of cell cycle process
```

The table can be filtered so that only the statistically significant results are retained:

```
go.table.sig<-go.table[go.table$Pvalue<=0.01,]
```

There are a total of 87 statistically significant GO categories. One of these is organelle fission which is also visible in the table above.

10.4.2 Self-Contained Method Using Globaltest Package

Globaltest package is rather straightforward to use, but it requires real expression data as input, and it needs to be converted into a suitable scale for the analysis. Package limma provides a way to do this using the function `voom()`, but it needs normalization factors and a desing matrix as inputs. Normalization factors can be calculated using the function `calcNormFactors()` from the package edgeR. Design matrix was generated earlier, when edgeR was used for detecting the significantly differentially expressed genes in this chapter. The preprocessing steps can be performed with the following code:

```
library(edgeR)
library(limma)
nf <- calcNormFactors(counts(parathyroidGenes))
y <- voom(counts(parathyroidGenes), design, plot=TRUE,
             lib.size=colSums(counts(parathyroidGenes))*nf)
```

The self-contained gene set analysis can then be performed using function `gtGO()`. This functions expect to get a matrix, where samples are on rows, and variables in the columns, so we can tell the function to do this automatically by setting an option:

```
library(globaltest)
gt.options(transpose=TRUE)
```

The actual analysis wants to get the response vector, which here is time, a matrix of expression values that were generated using `voom()`, the ontology to be tested against (here BP), and the name of annotation package that allows conversion of "probe names" into Entrez Gene IDs. Because we are using an organism-specific annotation package to perform the analysis, an additional parameter, probe2entrez is required. Here, it is a list of Ensembl gene IDs with their corresponding Entrez Gene IDs. The object ensemble.to.entrez was generated during the competitive analysis. The analysis is done with the following command:

```
go2<-gtGO(meta$time, y$E, ontology="BP",
      annotation="org.Hs.eg.db",
      probe2entrez=as.list(ensembl.to.entrez))
```

The result is a list with *p*-values for each GO category:

```
head(go2)
```

	holm	alias	p-value	Statistic
GO:0071168	1.94e-05	protein localization to chromatin	1.58e-09	55.9
GO:0051303	2.10e-05	establishment of chromosome localization	1.71e-09	40.7
GO:0050000	2.62e-05	chromosome localization	2.14e-09	37.1
GO:0046104	2.04e-04	thymidine metabolic process	1.67e-08	50.3
GO:0046125	2.04e-04	pyrimidined eoxyribo nucleoside metabolic process	1.67e-08	50.3
GO:0060138	2.07e-04	fetal process involved in parturition	1.69e-08	72.6

```
             Expected     Std.dev      #Cov
GO:0071168       3.85        3.92         7
GO:0051303       3.85        2.92        22
GO:0050000       3.85        2.70        23
GO:0046104       3.85        3.82         4
GO:0046125       3.85        3.82         4
GO:0060138       3.85        5.33         1
```

The table can be filtered to retain only the statistically significant GO categories:

```
go2.table.sig<-go2@result[go2@result[,1]<=0.01,]
```

The filtered table now holds results for 997 GO categories, which is about 10 times more than the number of categories returned by the competitive method.

10.4.3 Length Bias Corrected Method

Package goseq implements a method that corrects for the length bias in RNA-seq experiment while doing a gene set analysis. The input for the goseq analysis is a named vector of zeros and ones, where ones mark the genes that were differentially expressed. Such a vector can be easily constructed from the previous results acquired previously in this chapter, but note that no duplicate gene name entries are allowed in the vector:

```
gene.vector<-as.numeric(reference.genes %in%
                              selected.genes)
names(gene.vector)<-reference.genes
```

Object `gene.vector` now contains the needed information. The analysis is done in three steps. First, a weighting for each gene is calculated using the function `nullp()`. In addition to the named gene vector, it takes two parameters, the genome version and the type of the gene identifier. Available ones can be checked using the functions `supportedGenomes()` and `supportedGeneIDs()`. After having been estimated, the actual test can be performed using the function `goseq()`. The testing has been restricted to GO ontology biological process only using the argument `test.cats`. Finally, goseq does not automatically perform a multiple testing correction for the results, so we need to add a separate step for it. Function `p.adjust()` does the actual adjustment, using the Benjamini and Hochberg's false discovery rate (BH). The whole analysis should proceed as follows:

```
library(goseq)
pwf<-nullp(gene.vector,"hg19","knownGene")
GO.wall<-goseq(pwf,"hg19","knownGene",
                    test.cats=c("GO:BP"))
GO.wall$padj<-p.adjust(
                    GO.wall$over_represented_ pvalue,
                    method="BH")
```

Object GO.wall is now a simple data frame that contains results for all GO categories. There are 78 statistically significantly overrepresented categories, which can be extracted into a separate table with:

```
go3.table.sig<-GO.wall[GO.wall$padj<=0.01,]
```

10.5 SUMMARY

For updating the annotations of genes, Bioconductor offers not only organism-specific annotation packages, but also a direct connection to the international BioMart database collection. The annotations can be utilized in functional analyses of the results. Gene set analysis is an often applied technique that helps to make biological sense of the results. There are at least three major categories of gene set analysis methods, competitive (e.g., GOstats), self-contained (e.g., globaltest), and length bias correcting (e.g., goseq) methods. In addition, these methods can be categorized on the basis of the sampling unit (gene or sample).

REFERENCES

1. Goeman J. and Bühlmann P. Analyzing gene expression data in terms of gene sets: Methodological issues. *Bioinformatics* 23:980–987, 2005.
2. Khatri P. and Draghici S. Ontological analysis of gene expression data: Current tools, limitations, and open problems. *Bioinformatics* 21:3587–3595, 2005.
3. Maciejewski H. Gene set analysis methods: Statistical models and methodological differences. *Briefings Bioinform* 1–15, 2013. http://m.bib.oxfordjournals.org/content/early/2013/02/09/bib.bbt002.abstract.
4. Barry W.T., Nobel A.B., and Wright F.A. Significance analysis of functional categories in gene expression studies: A structured permutation approach. *Bioinformatics* 21(9):1943–1949, 2005.
5. Goeman J.J., van de Geer S.A., de Kort F., and van Houwelingen J.C. A global test for groups of genes: Testing association with a clinical outcome. *Bioinformatics* 20(1):93–99, 2004.

CHAPTER 11

Visualization

11.1 INTRODUCTION

Visualization is a general term covering everything from simple, exploratory plots to refined, publication quality graphs. Exploratory plots are often generated during the analysis to get to know the data or to check the output of the different analysis steps. Examples of such plots are histogram, scatterplot, and boxplot that give a simple and fast overview of the data set at hand. Publication quality graphs are something that are produced more meticulously. They are often grafted to highlight the results or to supplement the conclusions of the study and are of such quality that they can be printed to a book or to a journal article.

Most scientific journals have their own requirements for the image files they accept for publication. These are usually mentioned in the instructions for the authors. Typically, the journals accept images in at least TIFF or PDF format, but not universally all journals do so. In addition, they might have requirements for the color model used for the images and they might limit the resolution of the image.

In this chapter, we will cover a few graph types that could be used for conveying the messages of a study in a presentation or a publication. Exploratory plots are covered in the chapters that discuss the different analyses. Since a basic requirement of a publication quality graph is a suitable resolution and file type of an image, we will also discuss different solutions in R that get a user a long way toward fulfilling these requirements.

Let us first review the basic concepts of computer graphics, file types, resolution, and color models. After getting a basic understanding of these concepts, we will cover the principles of generating certain graphs and plots using R.

11.1.1 Image File Types

There are two broad categories of image file formats, bitmaps and vector graphics. Bitmap images, sometimes also called pixmaps or raster images, consist of individual points, pixels. Because the number of pixels in any given image is fixed, bitmap images cannot be enlarged without losing details in the process. In contrast, vector images use geometric shapes to represent images. Because vector images basically consist of mathematical expressions, their size can be changed without loss of quality. However, there is a slight caveat. If the number of elements, for example, points in a scatter plot, is large, vector graphic images will get very big, and files slow to open. Therefore, if the number of elements is large, it is best to use raster image formats to save the plots.

Examples of bitmap formats are Windows bitmap (BMP), tagged image file format (TIFF), joint photographic experts group (JPEG), and portable network graphics (PNG). The most often used vector graphic formats are PostScript and portable document format (PDF). To be exact, the (encapsulated) PostScript and the PDF are metafile formats that can incorporate both raster and vector images. However, if you generate a PostScript or a PDF file from R using the direction from this chapter, they will only contain vector images.

11.1.2 Image Resolution

Image resolution measures the detail of an image, and for bitmap images it is measured in pixels. For vector graphics, the resolution does not have a direct interpretation. For example, a TIFF image that contains 800 columns and 600 rows of pixels is said to have a resolution of 800×600, and it contains 480,000 individual pixels. In the digital camera world, this would be an image of (about) 0.5 megapixels.

In digital press, the resolution is often specified in points per inch (PPI) or dots per inch (DPI). It can be interpreted as the number of pixels per inch (25.4 mm). This information can be used for calculating the size of the image that you would need to produce. For example, if you are planning to print an image that is 4 inches (about 10 cm) wide, and the journal states that it needs to get images with a resolution of 600 DPI, you need to have an image of at least 2400 pixels wide in order to fulfill the requirement.

11.1.3 Color Models

The two most often utilized color models are RGB and CMYK. RGB is used on media that transfers light, such as television or a computer monitor. Each pixel of an image consists of three different colors, red (R), green (G), and blue (B), and when they mix together in specified quantities, they form the visible colors ("additive color format"). Since computer monitors work in the RGB mode, the images generated using computers are often saved into RGB-formatted files.

CMYK is used in printing industry. A large range of human visible colors can be formed by mixing inks of cyan (C), magenta (M), and yellow (Y). In addition, black (K) is often utilized for economic and technical reasons. CMYK is a subtractive color format, since inks absorb different lengths of light, and the visible color forms from the wavelengths that are not absorbed.

Some journals absolutely require you to submit the image in the CMYK format, while others accept images in the RGB format also.

11.2 GRAPHICS IN R

R can save images in several file formats. Among others, BMP, JPEG, TIFF, PNG, PDF, and PostScript are readily available. Both PostScript and PDF files can be generated in the RGB and the CMYK color formats, but all other formats support only the RGB color model. Resolution of the images can be readily customized and most resolution requirements are easily fulfilled.

There are two fundamentally different graphics systems in R, base graphics and grid graphics [1]. The base graphics contain both high-level and low-level functions. High-level functions produce a complete plot and low-level functions allow the user to generate the plots from the ground up, or to add something to an existing plot. Grid graphics do not contain high-level functions at all, but several packages, most notably lattice and ggplot2, use the grid graphics to implement a large variety of high-level plotting functions. Both base and grid graphics are static, and once generated, they cannot be modified by other means but plotting them again.

Whereas static graphics are available in infinite varieties, the same cannot be said about interactive graphics. There are a few add-on packages, such as iplots, playwith, and rgl, that add this kind of functionality to R, but the graph types available for interactive graphics are not very extensive. This limitation is solved to some extend by making R to interact with the ggobi program or by generating the graphs using some JavaScript libraries, such as rCharts. All static graphics can be saved into the aforementioned bitmap or vector graphics formats, but the same does not necessarily apply to the interactive graphics.

The following sections will cover the most common visualizations seen in the publications discussing gene expression and how to generate them in R.

11.2.1 Heatmap

Heatmap is a visualization that displays the expression values of the features (genes, exons, etc.) using a color scale. Features are typically arranged in columns (samples) and rows (features) as in the original data matrix. Each feature–sample pair is represented with a small rectangle that is colored according to its expression. Often both samples and features are hierarchically clustered before constructing the heatmap, and clustering is represented with a tree on the left and top from the colored data matrix.

Heatmap is often used to show that two or more groups have dissimilar expression of certain genes. This is fine if the plot is generated before any statistical testing has been performed. However, if the plot contains only the features (genes) that have been found to be statistically significantly differentially expressed, the heatmap cannot be used for showing that the two groups are really different. It can only work as a way to visualize the results in such a case. Let us illustrate both uses with examples using parathyroid data set in R.

The default function for generating heatmaps in R is `heatmap()`. It does not allow for very much customization of the plot, and here the function `pheatmap()` from the package pheatmap (pretty heatmap) is used instead. A pretty heatmap for the whole parathyroidGenesdataset can be generated roughly as follows:

```
# Loads the data
library(parathyroid)
data(parathyroidGenes)
# Filtering
keep <-rowSums(counts(parathyroidGenes) >100)
                  >=ncol(counts(parathyroidGenes))
dooku <-counts(parathyroidGenes)[keep,]
rsd <-rowSums(dooku)
dooku <-dooku[order(rsd),]
# Plotting
library(pheatmap)
pheatmap(log2(dooku),cluster_rows = FALSE,
            show_ rownames = FALSE,
            annotation = data.frame(
```

```
        (pData(parathyroidGenes)[,3:5])),
    border_color = "grey95",
    scale = "column")
```

First we load the data from the parathyroid package. Next, the data set is filtered to remove genes that have low counts in half of the samples. The count table is sorted according to the gene's total expression across samples to make the heatmap easier to read, and the counts are log 2-transformed. Finally, the plot is generated, and it should appear similar to Figure 11.1. Note that the columns are log 2-transformed and scaled to the same mean during plotting, and this can be thought of as a rough normalization, if a better solution is missing.

Alternatively, you can use the result from the DESeq analysis that was performed in Chapter 8. The following code will produce the heatmap using the variance-stabilized ("normalized") count data (the image is not shown):

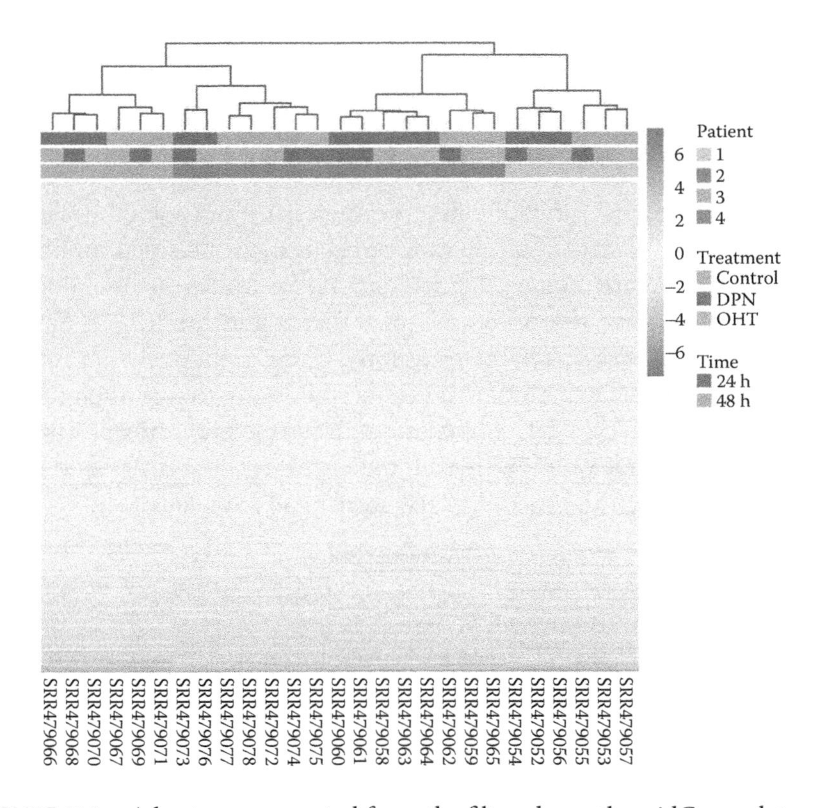

FIGURE 11.1 A heatmap generated from the filtered parathyroidGenes data set. By default a color blind-friendly color scheme ranging from red to blue is used.

```
library(parathyroid)
library(DESeq2)
data(parathyroidGenes)
d.deseq <- DESeqDataSetFromMatrix(
                countData = counts (parathyroidGenes),
                colData=pData(parathyroidGenes),
                design=~treatment)
d.deseq <-estimateSizeFactors(d.deseq)
d.deseq <-DESeq(d.deseq)
resultsNames(d.deseq)
res <- results(d.deseq,"treatment_OHT_vs_Control")
sig <- res[which(res$pvalue < 0.01),]
vsd <- getVarianceStabilizedData(d.deseq)
vsdp <- vsd[rownames(vsd) %in% rownames(sig),]
library(pheatmap)
pheatmap(vsdp, cluster_rows = TRUE,
           show_ rownames = TRUE,
           annotation = data.frame(
                (pData(parathyroidGenes)
                [,4,drop = FALSE])
           ),
           border_color = "grey95", scale = "none")
```

The pheatmap() function offers an interesting and sometimes a very useful feature. This plotting function can generate a specified number of pseudogenes that are then used as the data set for the image. Pseudogenes are generated using the K-means clustering algorithm which clusters together all the genes that behave similarly. Using pseudogenes in the plot also makes it possible to cluster the rows, something that is often impossible for the whole data set of tens of hundreds of thousands of features. The plot can be generated by setting the argument kmeans_k to some number (here, 100), and turning the clustering of rows on:

```
pheatmap(log2(dooku), cluster_rows = TRUE,
            show_ rownames = FALSE,
            annotation = data.frame(
                (pData(parathyroidGenes)
                [,3:5])
            ),
            border_color = "grey95",
            scale = "none", kmeans_k = 100)
```

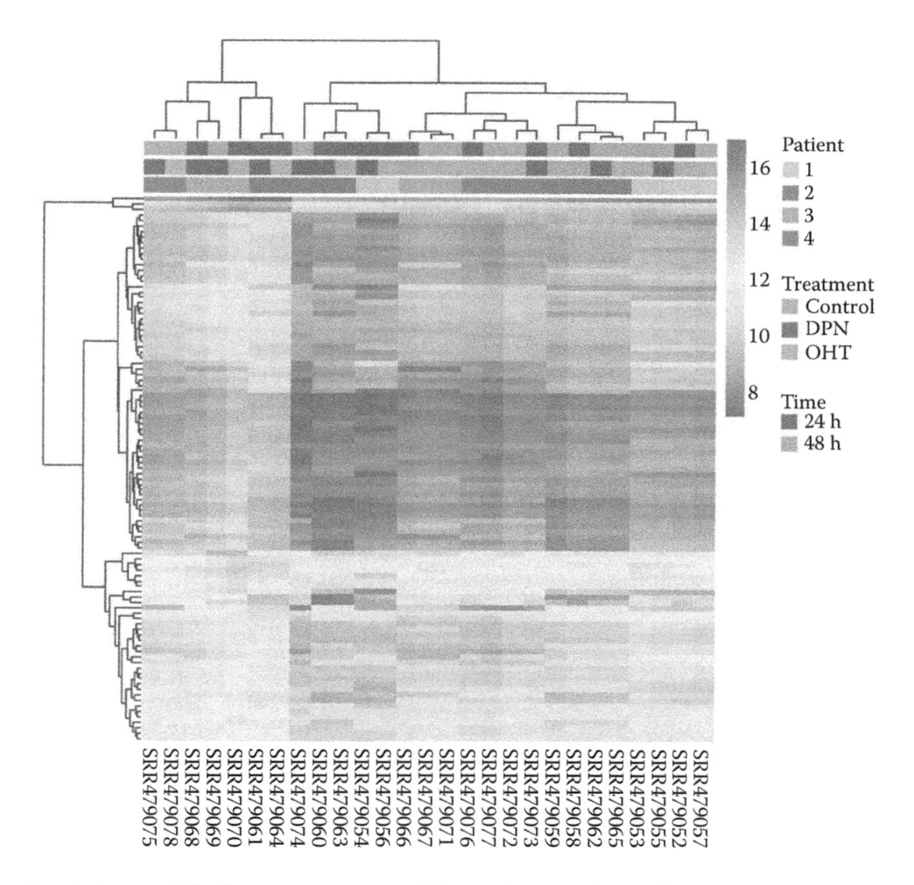

FIGURE 11.2 The heatmap generated from the parathyroidGenes data set using the pseudogenes for the plot.

The resulting plot is shown in Figure 11.2.

A heatmap can also be used for visualizing the results of the exon-specific expression analysis. Each exon forms a separate row in the plot, and the color is taken from the count table. For the gene ENSG00000119541 that was identified in Chapter 9, the heatmap can be generated from the object ecs (generated in Chapter 9) using the following R code:

```
vismat <-counts(ecs)[fData(ecs)$geneID==
                    "ENSG00000119541",]
colnames(vismat)<-pData(ecs)$samplename
pheatmap(log2(vismat), cluster_rows = FALSE,
          cluster_cols = FALSE, border_col = "grey95")
```

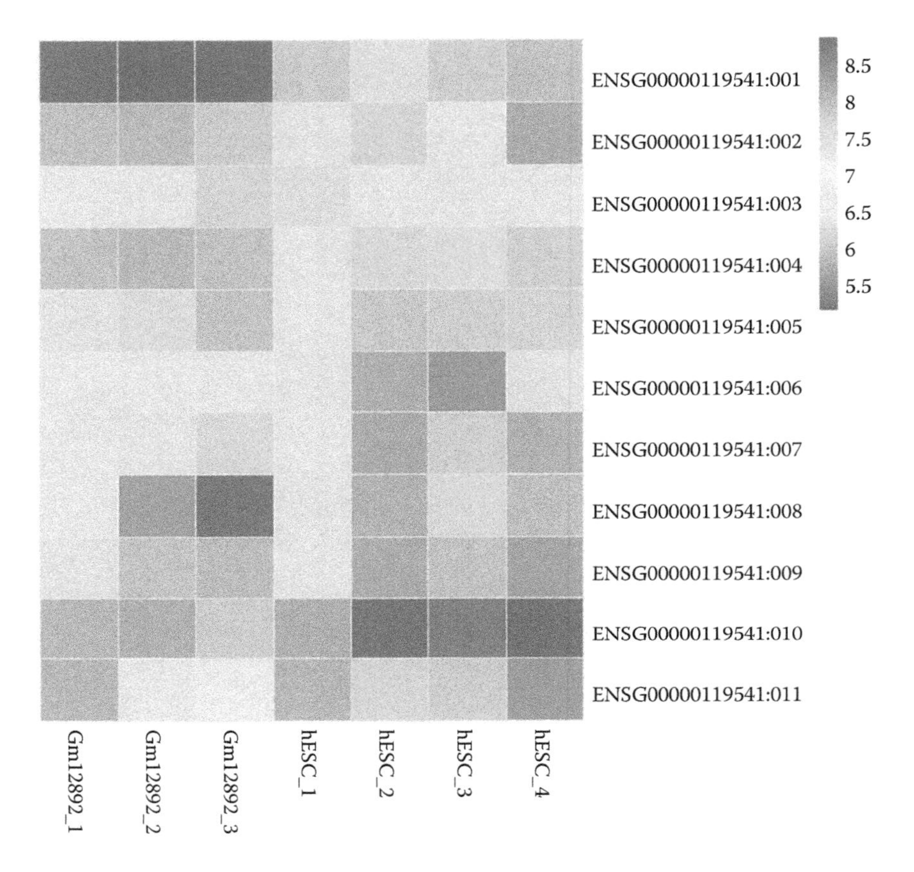

FIGURE 11.3 A heatmap generated from a count table for one gene's exons. Compare the plot with Figure 9.4.

First, all the exon-specific counts are extracted from the exonCountSet object (ecs), and the plot is generated without clustering the samples or the rows in order to retain the spatial arrangement of the exons along the gene. The resulting plot is shown in Figure 11.3.

11.2.2 Volcano Plot

A Volcano plot [2] is simply a scatterplot that has the fold change values for all features on the horizontal (x) axis, and the –log 10-transformed p-value on the vertical (y) axis. The plot is sometimes called a Volcano plot, because it resembles a volcano that is just erupting and splashing lava all over the place. The Volcano plot becomes nicer to look at, if the up- and downregulated features are colored differently, and if a couple of guiding lines are added to the background. A Volcano plot gives an easily

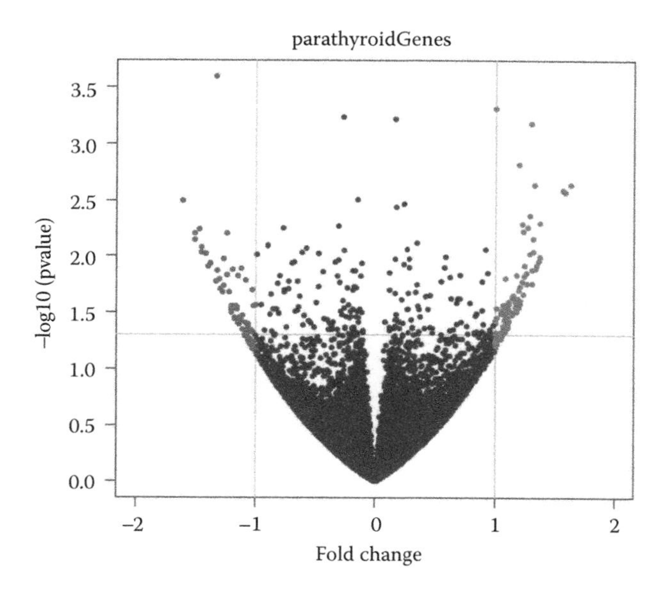

FIGURE 11.4 A Volcano plot generated from the parathyroidGenes data set. A rather low number of genes are statistically significantly differentially expressed, and this fact is clearly visible.

decodable overview of the rough number of under- and overexpressed features and their statistical significance.

Let us use the res object generated for the parathyroidGenes data set. The analysis is detailed in Chapter 8, but also above where the heatmap was discussed. The said res object contains columns log2FoldChange and pvalue that are used for a plot. The following code will produce the plot that is shown in Figure 11.4:

```
# Analysis library(parathyroid)
library(DESeq2)
data(parathyroidGenes)
d.deseq <- DESeqDataSetFromMatrix(countData =
                counts (parathyroidGenes),
                colData = pData(parathyroidGenes),
                design = ~treatment)
d.deseq <- estimateSizeFactors(d.deseq)
d.deseq <- DESeq(d.deseq)
resultsNames(d.deseq)
res <- results(d.deseq, "treatment_OHT_vs_Control")
sig <- res[which(res$pvalue < 0.01),]
```

```
# Generate the colors
cols <-rep("#000000", nrow(res))
cols[res$log2FoldChange >=1]<-"#CC0000"
cols[res$log2FoldChange <=-1]<-"#0000CC"
# Produce the plot
plot(res$log2FoldChange,-log10(res$pvalue),
   pch = 16, cex = 0.75, col = cols, las = 1,
   xlab = "FoldChange", ylab = "-log10(pvalue)",
   xlim = c(-2,2))
# Add the vertical and horizontal
lines abline(h = -log10(c(0.05)),col = "grey75")
abline(v = c(-1,1),col = "grey75")
# Plot the points again to overlay the lines
points(res$log2FoldChange, -log10(res$pvalue),
   pch = 16,cex = 0.75, col = cols)
# Add a title
title(main = "parathyroidGenes")
```

11.2.3 MA Plot

MA plot [3] is a scatter plot where an average expression of a gene is put on the x-axis, and the fold change or log ratio is on the y-axis. MA plot has been extensively used in the field of DNA microarray bioinformatics. It is also usable with RNA-seq experiments. Here is an example of the parathyroidGenes data set. First, let us find the genes that differentially expressed:

```
# Analysis library(parathyroid)
library(DESeq2)
data(parathyroidGenes)
d.deseq <- DESeqDataSetFromMatrix(countData =
   counts (parathyroidGenes),
   colData = pData(parathyroidGenes),
   design = ~treatment)
d.deseq <- estimateSizeFactors(d.deseq)
d.deseq <- DESeq(d.deseq)
resultsNames(d.deseq)
res <-results(d.deseq, "treatment_OHT_vs_Control")
sig <- res[which(res$pvalue < 0.01),]
```

The resulting object res contains the columns baseMean (average expression) and the column log2FoldChange that contains the fold change values. In addition, the column pvalue contains the raw p-values that are

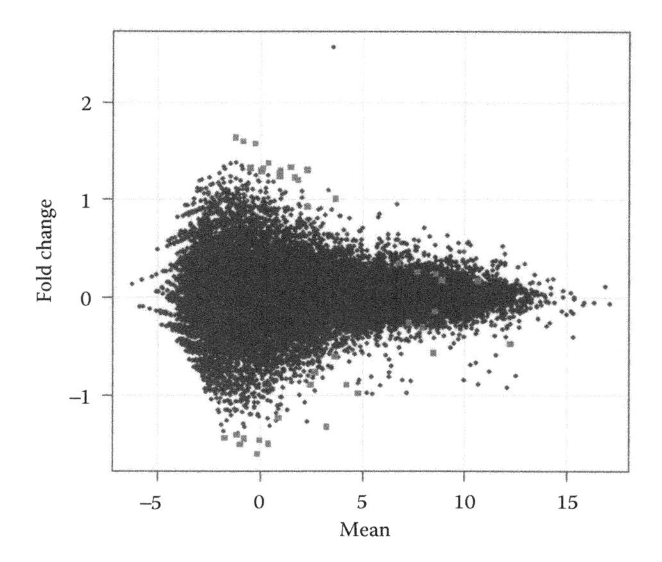

FIGURE 11.5 An example of an MA plot for the parathyroidGenes data set. The expressed genes are highlighted with the squares.

used for coloring the symbols in the plot. First, we plot the scatterplot, and a gray grid over it. Then to erase the gray lines of the grid that overlay the points, we plot the points again, but separately for the differentially expressed and unexpressed genes. Differentially expressed genes are plotted last with red squares so that they are highlighted relative to the unexpressed genes. Last, the box that surrounds the plot area is reinforced to give the plot a finalized appearance. The R code for plot is printed below, and the sample image is shown in Figure 11.5.

```
# Plot
plot(x = log2(res$baseMean), res$log2FoldChange, pch = 16,
   cex = 0.5, las = 1, xlab = "Mean", ylab = "Fold Change")
grid(lty = 1, col = "grey95")
cols <-ifelse(res$pvalue>0.01|is.na(res$pvalue),
   "#000000", "#CC0000")
points(x = log2(res$baseMean)[cols = ="#000000"],
   res$log2FoldChange[cols = ="#000000"],pch = 16,
   cex = 0.5)
points(x = log2(res$baseMean)[cols = ="#CC0000"],
   res$log2FoldChange[cols = ="#CC0000"],pch = 15,
   cex = 0.75, col = "#CC0000")
box()
```

11.2.4 Idiogram

Idiogram is a drawing or a photograph of the chromosomes (karyotype) of a cell. It is more generally used to depict an idealized graph of the organism's chromosomes, something that is occasionally also called, slightly misleadingly, an ideogram. In gene expression studies, it is sometimes used for displaying the locations of the expressed genes in the organism's genome. In addition, locations of SNPs and structural variations are similarly often visualized using an idiogram.

Human chromosomes appear striped under the microscope after Giemsa staining. These stripes are called isochores, which are large areas of genome that have a different relative GC content. Isochores are often used for indicating the locations of the genes in the chromosomes. Before the human genome was mostly sequenced, isochores were the dominant positioning system. Nowadays, exact nucleotide positions are more often used instead. The slight downside of the nucleotide-based location system is that the positions might change between different assembly versions of the same genome. The isochores are relatively more stable, but the amount of the stripes somewhat depends on the cell-cycle stage the chromosomes were stained. Using the isochores (bands), the gene locations are indicated using a simple system of chromosome number (1–22, X, and Y), chromosome arm (q for long arm and p for short arm), and isochore band (running numbering from the centromere toward the ends of the chromosome). For example, 7q31 would indicate that the gene is located in the chromosome 7's long arm, band 31. Subbands are also sometimes used, and are indicated using decimals, for example, 7q31.2.

Let us generate an idiogram that indicates the positions of the differentially expressed genes found from the parathyroidGenes data set. First, we need to find what the differentially expressed genes are. This is done as above for the heatmap, but the process is repeated here:

```
library(parathyroid)
library(parathyroid)
library(DESeq2)
data(parathyroidGenes)
d.deseq <- DESeqDataSetFromMatrix(
   countData = counts(parathyroidGenes),
   colData = pData(parathyroidGenes),
   design = ~treatment)
d.deseq <- estimateSizeFactors(d.deseq)
```

```
d.deseq <- DESeq(d.deseq)
resultsNames(d.deseq)
res <- results(d.deseq, "treatment_OHT_vs_Control")
sig <- res[which(res$pvalue < 0.01),]
```

After the significantly differentially expressed genes are saved into the object sig, we need to find the chromosomal locations for the genes. There are several possibilities for this, but here an organism-specific annotation package for human is used.

```
library(org.Hs.eg.db)
keys <- keys(org.Hs.eg.db, keytype = "ENSEMBL")
columns <-c ("CHR","CHRLOC", "CHRLOCEND")
sel <-select(org.Hs.eg.db, keys, columns,
   keytype = "ENSEMBL")
sel2 <-sel[sel$ENSEMBL %in% rownames(sig),]
sel3 <-na.omit(sel2[!duplicated(sel2$ENSEMBL),])
sel3$strand <-ifelse(sel3$CHRLOC < 0, "-", " + ")
sel3$start <-abs(sel3$CHRLOC)
sel3$end <-abs(sel3$CHRLOCEND)
```

Once the locations of the differentially expressed genes are known, we can use the data for generating an idiogram. There are actually a few possibilities for plotting an ideogram, and Bioconductor project offers at least two packages that offer this functionality. The older package is idiogram that works nicely with the package `geneplotter`. However, here we will use a newer package ggbio. Before the actual plotting, we need to generate a Granges object that contains the location of the differentially expressed genes, because ggbio works mainly with Granges objects:

```
library(GenomicRanges)
tt <-org.Hs.egCHRLENGTHS
read<-GRanges(
      seqnames=Rle(paste("chr", sel3$CHR, sep="")),
      ranges=IRanges(start=sel3$start, end=sel3$end),
      strand=Rle(sel3$strand)
)
```

After the Granges object is in place, an idiogram is produced as follows. First, we need the data on the banding of the human chromosomes.

This is contained in the package biovizBase in the object hg19Ideogram-Cyto. Note that this data go hand in hand with the genome assembly version hg19. Since we want only the complete chromosomes (there also other fragments in the data), we will select only those with the function `keepSeqlevels()`. The idiogram is then generated with the function `ggplot()` using a karyogram-specific layout. The gene locations are then added as red bars to the plot to form the final idiogram. The plot is rendered when the object storing it is printed simply by typing its name (p):

```
library(biovizBase)
data(hg19IdeogramCyto, package = "biovizBase")
data(hg19Ideogram, package = "biovizBase")
hg19 <- keepSeqlevels(hg19IdeogramCyto,
   paste0 ("chr", c(1:22, "X", "Y"))) library(ggbio)
p <- ggplot(hg19) + layout_karyogram(cytoband = TRUE)
seqlengths(read) <-
   seqlengths(hg19Ideogram)[names(seqlengths(read))]
p <- p + layout_karyogram(read, geom = "rect",
   ylim = c(11,21), color = "red")
p
```

The resulting plot is shown in Figure 11.6.

11.2.5 Visualizing Gene and Transcript Structures

Known transcripts are rather straightforward to visualize using the ggbio package. Known human transcripts for the human genome assembly hg19 are available as a Bioconductor package. A plot containing the gene structure (exons and introns), and the strand information (forward or reverse) and the chromosomal location is generated in several steps.

First all known transcripts are extracted from the Bioconductor package `TxDb.Hsapiens.UCSC.hg19.knownGene`. The start and end positions (in base pairs) are then taken from the organism-specific annotation package `org.Hs.eg.db`. The location is then converted into a Granges object as above for the idiogram and plotted with the function `autoplot()`. Each transcript is plotted on a separate line, but the gene structure can also be reduced. A reduced version plots all the transcripts on top of each other, which contains all the exons and introns of the gene on a single line. The chromosomal position can be plotted with the function `plotIdeogram()`. Finally, a complete plot containing all the elements

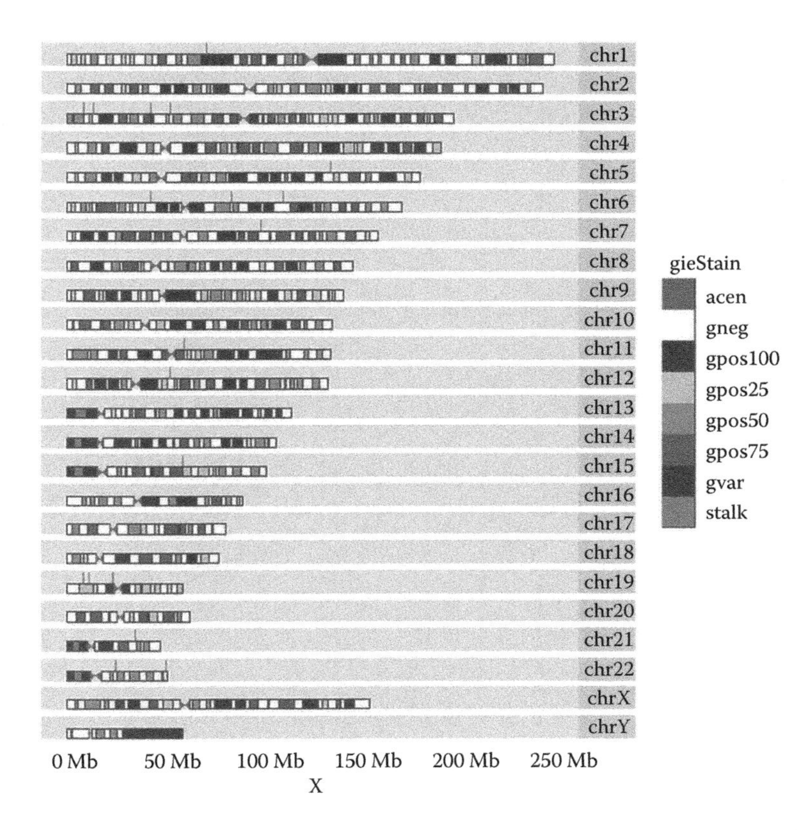

FIGURE 11.6 An idiogram of the human chromosomes with the positions of the differentially expressed genes inferred from the parathyroidGenes data set overlaid.

can be generated using the function `tracks()`. The resulting plot is shown in Figure 11.7, and the whole R code is listed below:

```
library(TxDb.Hsapiens.UCSC.hg19.knownGene)
library(org.Hs.eg.db) library(ggbio)
txdb <- TxDb.Hsapiens.UCSC.hg19.knownGene
columns <-c ("CHR", "CHRLOC", "CHRLOCEND")
sel <-select(org.Hs.eg.db, "ENSG00000119541", columns,
   keytype = "ENSEMBL")
wh <- GRanges("chr18", IRanges(61056425, 61089752),
   strand = Rle("-"))
p1 <- autoplot(txdb, which = wh, names.expr = "gene_id")
p2 <- autoplot(txdb, which = wh, stat = "reduce",
   color = "brown",fill = "brown")
```

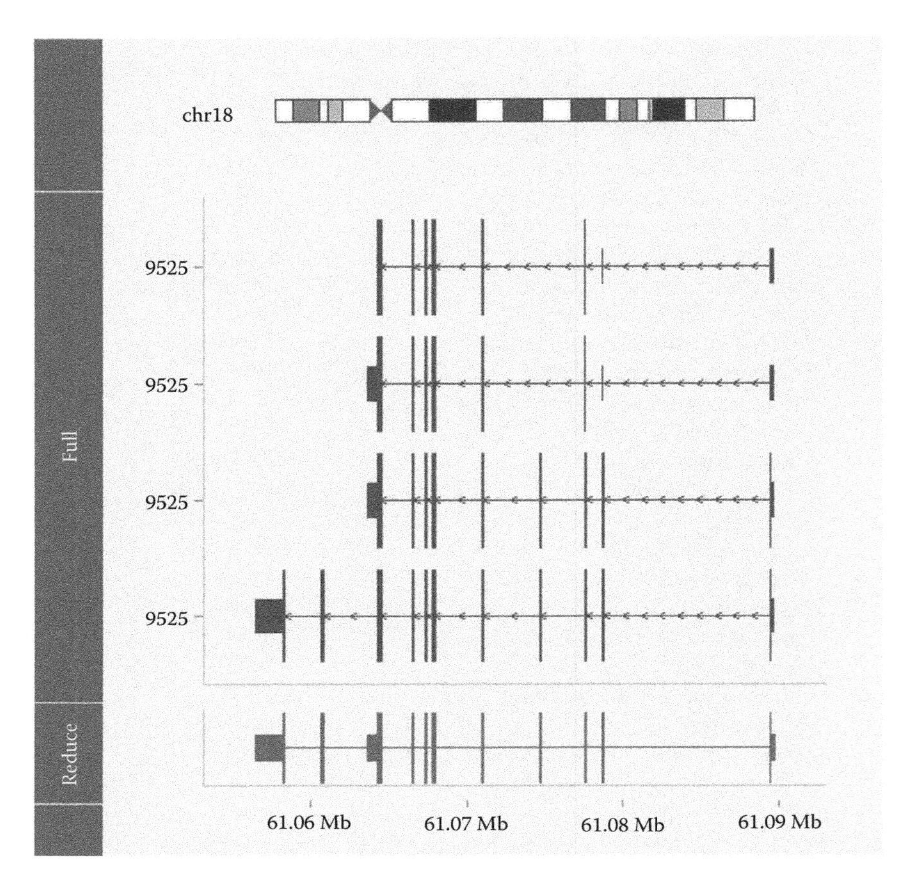

FIGURE 11.7 A plot of the gene ("reduce") and transcript ("full") structures for the gene ENSG00000119541 that was found to have two differentially expressed exons in the ENCODE data set. The gene is located in the long arm of the chromosome 18.

```
p.ideo <- plotIdeogram(genome = "hg19", subchr = "chr18")
tracks(p.ideo, full = p1, reduce = p2,
   heights = c(1.5, 5, 1)) + ylab("") +
   theme_tracks_sunset()
```

11.3 FINALIZING THE PLOTS

All the codes above produce the plot on screen. It is possible to save the plot to an image file on Windows and Mac OS via menu selections, but it is much better to take better control over the details of the produced plot. As mentioned in the chapter's introduction, it is possible to change the resolution of the raster or bitmap images, and additionally for the vector graphics, the color model can be specified.

First of all, the device to which we want to plot must be opened before the actual image is plotted. There are several graphics devices that can be used. If the aim is to produce a publication quality plot, the TIFF or PDF formats are probably the most universally accepted ones. The graphics devices are opened for these file types by the command `tiff()` or `pdf()`. After the graphics device is opened, the individual plotting commands are given as in the examples above. After plotting the commands, the graphic device must be closed, which is done with the command `dev.off()`. It is absolutely vital to remember to close the graphics device. Otherwise, all your further screen outputs will also be directed to the graphics device and to the plot.

For example, if we want to plot the results for a single gene from the exon-wise analyses as heatmap, we use the following code that will generate a tiff image:

```
tiff(filename = "heatmap1.tif", width = 1000, height =
1000)
   vismat <-counts(ecs)
      [fData(ecs)$geneID = ="ENSG00000119541",]
   colnames(vismat) <-pData(ecs)$samplename
   pheatmap(log2(vismat), cluster_rows = FALSE,
      cluster_cols = FALSE, border_col = "grey95")
dev.off()
```

The code will produce a tiff image that is 1000 pixels high and wide. On paper, this will typically be a little more than three inches (8 cm) by side. If you test the code above, you will probably find out that the text in the plot has become very small. Therefore, you will need to adjust the plot margins and the text size. These can be performed by setting the usual R graphic parameters via the function `par()` or inside the plotting command, for example,

```
tiff(filename = "heatmap1.tif", width = 1000, height =
1000)
   vismat <-counts(ecs)
      [fData(ecs)$geneID = ="ENSG00000119541",]
   colnames(vismat) <-pData(ecs)$samplename
   par(mar = c(4,1,1,4))
   pheatmap(log2(vismat), cluster_rows = FALSE,
```

```
      cluster_cols = FALSE, border_col = "grey95",
      cex = 1.5)
dev.off()
```

For a pdf graphics device, you do not specify the width and height as pixels, but in inches instead. In principle, you can still use pixels, but remember to divide them by the resolution, or you will end up with a huge file. For a printing press that want to get the file in 300 DPI resolution, you can divide the amount of pixels by 300, for example,

```
pdf(file = "heatmap1.pdf", width = 1000/300,
   height = 1000/300)
      vismat <-counts(ecs)
         [fData(ecs)$geneID = ="ENSG00000119541",]
      colnames(vismat) <-pData(ecs)$samplename
      pheatmap(log2(vismat), cluster_rows = FALSE,
        cluster_cols = FALSE,
        border_col = "grey95",
        cex = 1)
dev.off()
```

Additionally, it is possible to generate a PDF file in CMYK color model, if needed. This requires one extra argument, color model that takes on value "cmyk":

```
pdf(file = "heatmap1.pdf", width = 1000/300,
   height = 1000/300, colormodel="cmyk")
      vismat <-counts(ecs)
         [fData(ecs)$geneID = ="ENSG00000119541",]
      colnames(vismat) <-pData(ecs)$samplename
      pheatmap(log2(vismat), cluster_rows = FALSE,
         cluster_cols = FALSE,
         border_col = "grey95",
         cex = 1)
dev.off()
```

11.4 SUMMARY

R can generate a multitude of different plots, such as heatmap, Volcano plot, MA plot, and idiogram. Also, more specialized plots that allow visualization of gene and transcript structures can be generated using R. In addition, R has functionality for saving the plots in various formats, such as Postscript, PDF, and TIFF, but different color models are only supported by vector graphics formats.

REFERENCES

1. Murrell P. *R Graphics*, Boca Raton: CRC Press, 2011, ISBN9781439831762.
2. Cui X. and Churchill G.A. Statistical tests for differential expression in cDNA microarray experiments. *Genome Biol* 4(4):210, 2003. Epub Mar 17, 2003 [Review]. PubMed PMID: 12702200; PubMed Central PMCID: PMC154570.
3. Yang Y.H., Dudoit S., Luu P., Lin D.M., Peng V., Ngai J., and Speed T.P. Normalization for cDNA microarray data: A robust composite method addressing single and multiple slide systematic variation. *Nucleic Acids Res* 30(4):e15, 2002. PubMed PMID: 11842121; PubMed Central PMCID: PMC100354.

CHAPTER 12

Small Noncoding RNAs

12.1 INTRODUCTION

Noncoding small RNAs are composed of multiple classes of biologically active molecules that control and modify biological functions ranging from early developmental timing to programmed cell death. They are expressed from before the first cell division to the final stages of aging. They can be found in oocytes, stem cells, neurons, glial cells, somatic cells, and also in cancer cells. Although early studies have focused on the identification and elucidation of the function of small RNAs in translation (tRNAs, snoRNAs), more recent studies have placed an emphasis of noncoding RNAs that provide control or modify transcription (miRNAs, piRNAs, endo-siRNAs). Early studies have also relied heavily on traditional molecular biology methods such as cloning and dideoxy sequencing to identify class members. More recent RNA-seq methodologies now provide unprecedented breadth and depth for sequencing small-RNA class members. With the tremendous advancements in throughput and accuracy of next-generation sequencing methodologies, the challenge now becomes identifying, annotating, and profiling the known small RNAs and discovering the novel RNAs within a sequencing data set. Current small-RNA-seq methods do not easily discriminate between small-RNA classes, therefore knowledge of the different classes, their characteristics, and relation to one another is beneficial for planning and performing experiments aimed at identifying and quantifying members of a particular small noncoding RNA class. Below, we describe the major small noncoding RNA classes. In the descriptions, we emphasize animal systems. A summary of the different classes is presented in Table 12.1. As the amount of small-RNA sequencing data available explodes upon us, it is becoming obvious that many different classes on noncoding RNA exist with new classes in the process of discovery. Below, we attempt to classify at least those that are well known and better characterized. The following list is not intended to be an exhaustive list of all small noncoding RNA classes, but a reasonable starting point in understanding the currently known main classes.

TABLE 12.1 Major Classes of Small Noncoding RNAs

Class	Size	Biogenesis	Number of Members (Humans)	Function
microRNA (miRNA)	21–23 nt	Processed by DICER from 65 to 70 nt precursors	>2500	Regulation of gene expression
piwi-RNA (piRNA)	25–33 nt	Nuclear precursor amplified by ping pong mechanism in cytoplasm	>20,000	Regulation of retro transposons
Endogenous silencing RNA (endo-siRNA)	21–26 nt	Processed from messenger RNA transcripts	Unknown	Regulation of gene expression
Small nucleolar RNA (snoRNA)	60–300 nt	Processed from messenger RNA introns	>260	Participates in chemical modification of other RNAs
Small nuclear RNA (snRNA)	150 nt	RNA polymerase II and III	9 families	Participates in splicing of other RNAs
Transfer RNA (tRNA)	73–93 nt	RNA polymerase III	>500	Translation of mRNA to protein
microRNA off-setRNA (moRNA)	19–23 nt	Processed from miRNA precursor	Unknown	Unknown
Enhancer RNA (eRNA)	50–2000 nt	Nascent RNA transcription	>2000	Regulate proximal gene expression

12.2 MICRORNAs (miRNAs)

MicroRNAs (miRNAs) are small 21–23 nt molecules processed from larger primary-miRNA and subsequent ~70 nt precursor-miRNA molecules. They are diverse, abundant, and evolutionarily conserved. The founding member, *lin-4*, was found serendipitously in forward genetic screens aimed at identifying genes whose mutation caused lineage defects in the development of *Caenorhabditis elegans* [1]. *lin-4* mutant animals belong to a class of heterochronic mutants that fail to develop adult structures such as the vulva due to developmental timing defects. The wild-type *lin-4* gene that rescued these defects in mutant animals was surprisingly found to not encode a protein, but a 22 nt small RNA that was processed from a small hairpin precursor. The *lin-4* small-RNA gene was also found to have strong sequence complementary to multiple sites of the 3′ untranslated region of *lin-14*. As the expression profile of *lin-4* and *lin-14* was inversely correlated, and *lin-4* was shown to be a repressor of *lin-14*, it was hypothesized that *lin-4* targets *lin-14* and ultimately causes down regulation of *lin-14* protein via recognition and then translational repression. The estimated size of the *lin-4* transcript was originally proposed to be approximately 22 nt in length with a longer putative precursor 61 nt in length. Since these early reports, very little was published in the field until another *C. elegans* cell lineage defect gene, *let-7*, was also cloned and identified as a 22 nt noncoding RNA. The subsequent identification of orthologs of *let-7* in humans, *Drosophila*, and other higher organisms led to an explosion of interest and efforts in the field and simultaneous cloning and identification of hundreds of novel miRNAs [2–5]. miRNAs have now been found to be conserved in many other species including humans, plants, and other nematodes. Currently, there are more than 24,000 entries in miRBase.

A great deal of effort has been made to uncover the biogenesis mechanism of miRNAs. miRNA genes are initially transcribed by RNA polymerase II, the same polymerase that produces mRNAs. Some may also be transcribed by RNA polymerase III. The miRNA gene that is initially transcribed, primary miRNA or pri-miRNA, may be located intergenically between protein-coding genes, within an intron, in the coding region, or in a nontranslated region of a mRNA. Many miRNAs are also transcribed from their own promoters. The range in size for the primary miRNA is large spanning from a few hundred nt up to several kbp. The next step is to generate a mature 60–70 nt hairpin intermediate called the precursor miRNA or pre-miRNA. This maturation step is performed by DroshaRNase III endonuclease in a large ~650 kDa complex with (DiGeorge syndrome critical region gene 8) DGCR8/Pasha, which contains two double-stranded RNA (dsRNA)-binding domains. This complex is often called the pri-miRNA processing complex or microprocessor complex. The result of processing the pri-miRNA is a stem loop with a 2 nt 3′ overhang, now termed the pre-miRNA. This pre-miRNA is then transported from the nucleus to the cytoplasm by Ran-GTP and Exportin-5. In the cytoplasm, Dicer, another RNAase III endonuclease recognizes the double-stranded pre-miRNA and cuts both strands of the RNA. Dicer cleaves away both the terminal 5′ and 3′ overhangs and cuts the loop, leaving a double-stranded 22 nt RNA with a 2 nt 3′ overhang in both strands of the duplex. Of the two strands, the active "guide" molecule is then loaded into an Argonaute protein in the RNAi silencing complex also known as RISC complex, while the other "passenger" strand is degraded. The RISC complex, which contains several other proteins including GW182, also known as AIN-1 in *C. elegans*, then locates and targets mRNA. The guide molecule brings the RISC complex to close proximity to the target and recruits enzymes and cofactors that then effects RNA silencing via translational repression, loss of mRNA stability, mRNA degradation, or a combination of all three. An overview of the distinct steps in miRNA processing is shown in Figure 12.1.

In Figure 12.1, miRNA genes are transcribed in the nucleus by RNA polymerase II to produce pri-miRNAs. These molecules are then cropped by Drosha complexed with DGCR8, an RNAase III endonuclease in the pri-miRNA processing complex to produce the pre-miRNA hairpin. The hairpin is exported to the cytoplasm by Exportin 5-RanGTP. In the cytoplasm, the hairpin is further processed by Dicer, another RNAase III endonuclease in the miRNA loading complex (miRLC), to produce the mature miRNA guide strand and passenger strand. The guide strand is then loaded into an Argonaute protein that is associated with GW182. Other associated proteins depend upon the RNA-induced silencing complex (RISC) pathway. In the translation repression pathway, poly-A binding protein, PABP, GW182, and proteins of the translational machinery interact to block translation. In the mRNA decay pathway, CCR4:NOT complex, which contains at least five CCR or NOT proteins, acts to deadenylate polyadenylated mRNAs, while mRNA-decapping enzyme 1/2, DCP1/2 may act later to decap m7G mRNAs. A model has also been proposed where translational repression followed by mRNA decay may occur sequentially [6].

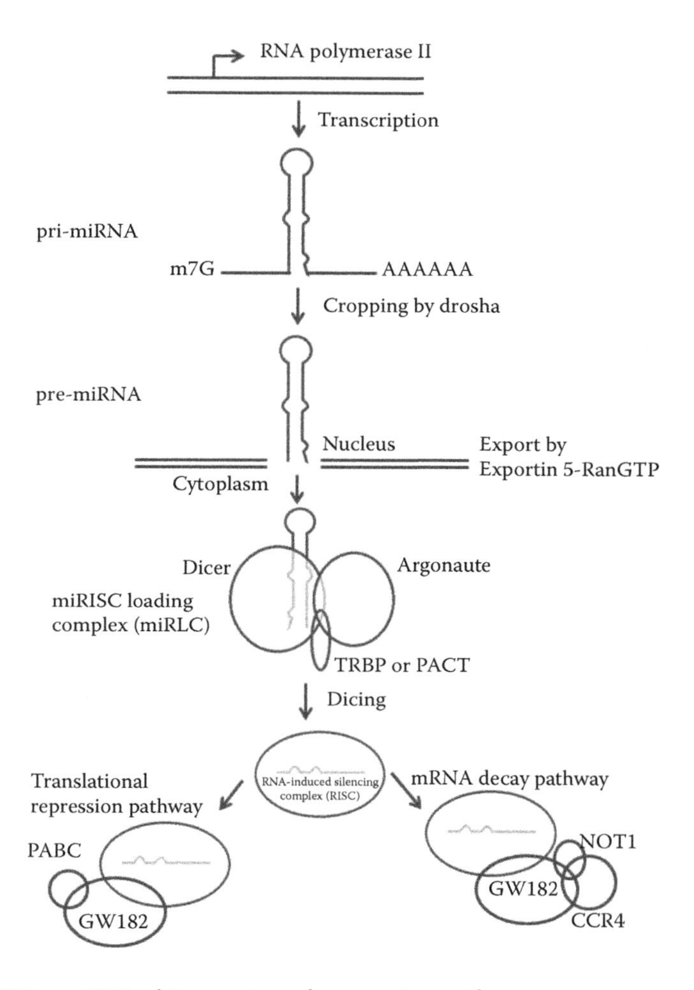

FIGURE 12.1 miRNA biogenesis and processing pathway.

Owing to the large number of miRNAs identified, overlap of their sequences, and their phylogenetic conservation, a specific nomenclature has been developed. The nomenclature description is shown in Table 12.2.

IsomiRs are isoforms of mature miRNAs and differ from them by a few nt in 5′ or 3′ directions, likely due to imprecision in processing during miRNA biogenesis [7]. They can also be generated by editing which may add one or more uridine residues to the 3′ end of a mature miRNA, or adenosine to inosine editing at the end of double-stranded RNA molecules by the adenosine deaminase enzyme. Specific isomiRs have been shown to be abundant, regulated, conserved, and functional, so therefore are considered biologically significant.

TABLE 12.2 miRNA Nomenclature

Term	Definition	Example
Species identifier	Abbreviated three or four letter prefixes	hsa-miR-101 (*Homo sapiens*), mus-miR-101 (*Mus musculus*), dme-miR-101 (*Drosophila melanogaster*), cbr-miR-101 (*Caenorhabditis briggsae*)
Mature identifier	Mature sequences processed by dicer are designated with upper case R	hsa-miR-7
Precursor identifier	Precursor hairpins are lower case R	hsa-mir-7
Numerical identifier	Sequential numbers are assigned based on historical precedent (earlier discovered miRNAs have lower numbers and more recently discovered have higher numbers). Orthologs have the same number in different species	mir-1, mir-2, mir-3
Paralog identifier	Paralogs that differ in the mature sequence by 1–2 nt are given lettered suffixes following the numerical identifier	mmu-miR-10a and mmu-miR-10b are mouse paralogs of mir-10
Distinct hairpin precursors, same mature miRNA sequence	Numbered suffixes	dme-mir-281–1 and dme-mir-281–2
Major form and minor form	Major more highly expressed form is named as described above, and the minor form with the asterisk "*" or star designation	mmu-miR-124 for the major form and mmu-miR-124* as the minor form
5 Prime arm and 3 prime arm	The star/* form have been replaced by a more unambiguous system where the mature miRNA in the 5 prime arm of the precursor is designated 5p and the mature miRNA from the 3 prime arm is designated 3p	miR-124-5p, miR-124-3p
miRNA cluster	Multiple miRNAs transcribed on a single RNA, or multiple miRNA genes located closely on a chromosome	miR-17-92 contains a single transcript from which six miRNA genes and 12 mature miRNAs are processed: miR-17, miR-18a, miR-19a, miR-20a, miR-19b-1,miR-92a-1
miRNA families	Defined by matching seed sequences	*let-7* family which contains in vertebrates *let-7*, mir-98, and mir-202

miRNAs can also be alternatively processed from intronic sequences via splicing and debranching via lariat debranching enzyme after which they are folded into pre-miRNA hairpins in the Mirtron pathway. As short hairpins derived from spliced introns, they bypass the microprocessor complex and are cleaved into mature miRNAs by Dicer in the cytoplasm after which they can proceed in the miRNA processing pathway as with other miRNAs [8]. Mirtrons were originally identified and cloned from *D. melanogaster* and *C. elegans*, as these have compact genomes and thus many short hairpin length introns, and later have been found in a large variety of species including humans, other primates, and plants.

12.3 MICRORNA OFF-SET RNAs (moRNAs)

MicroRNA off-set RNAs (moRNAs) are small-RNA molecules derived from the same pre-miRNA hairpin as miRNAs. moRNAs are located adjacent to mature miRNA in both the 5′ and/or 3′ arm and have a similar size as mature miRNAs. They have been identified in the simple chordate *Ciona instestinalis* and later in human brain libraries through RNA-seq analysis [9–11]. Although far less abundant than miRNAs, the function and details of its processing through the hairpin are at present unknown.

12.4 PIWI-ASSOCIATED RNAs (piRNAs)

Piwi-associated RNAs (piRNAs) are also small noncoding RNAs but have unique expression patterns, biogenesis, and function. piRNAs are approximately 25–33 nt long, expressed in germ line cells, especially in testes but can also be found in female cells and defined by their association with PIWI proteins, a class of Argonautes. The sequences also have a bias toward U in the first position. They function to silence mobile germline genomic DNA sequences, commonly referred to as transposable elements. Transposable elements are mobile sequences of DNA in a genome. The transposable elements can be classified as "retrotransposons," which require their transcription to RNA, reverse transcription to DNA, and then reinsertion into the genome; or "DNA transposon," which requires an active transposase enzyme gene, whose product catalyzes the excision and integration of a DNA transposable elements. The most common transposable element in the human genome is the Alu sequence which is approximately 300 bp long and present in several hundred thousand copies. piRNAs are initially transcribed as primary piRNAs from specific regions of the chromosome that can be from a few kbp to greater than 200 kbp. These regions are termed piRNA clusters and each cluster can account for tens or thousands of piRNA sequences. The piRNA sequences from each cluster can overlap and can also be transcribed from both strands within the cluster. Primary piRNAs are then loaded into one of three Argonautes: MILI, MIWII, or MIWIII in mice; and PIWI, AUB, or AGO3 in *D. melanogaster.* It should be noted that each Argonaute has size specificity. For example, in *D. melanogaster*, the piRNA sizes for the respected Argonautes are 25, 24, or 23 nt for PIWI, AUB, and AGO3, respectively. Primary piRNAs then undergo a unique second biogenesis step in which they are amplified. Sense piRNAs loaded to AUB bind to complementary antisense strands and are cleaved 10 nucleotides from the 5′ end. The secondary piRNA is then loaded onto AGO3 and binds the sense strand of an active transposon transcript and is cut 10 nt from the 5′ end, thus regenerating a piRNA from the transposon sequence. The cutting activity is termed slicer and contained within the AUB and AGO3. This amplification loop has been called the "ping-pong amplification cycle."

12.5 ENDOGENOUS SILENCING RNAs (endo-siRNAs)

Endogenous silencing RNAs (endo-siRNAs) are endogenously transcribed small RNAs 21 nt in length. They are generated from double-stranded RNA products that can arise from transcription of sense–antisense short pairs, single RNA strand inverted repeats, or hybridization of antisense pseudogenes with sense-coding genes. The sense–antisense pairs can be generated from transcription of a single locus in both orientations, termed *cis*-dsRNA, or from distinct loci with complementary sequences, termed *trans*-dsRNA. Sense–antisense precursors can also be generated from hairpins, but differ from miRNA hairpins by having longer stems and also by processing by DICER2 instead of DICER1. The complementarity between *trans*-dsRNA and long stem hairpins may not be exact resulting in many bulges in the dsRNA and may also be edited. The primary endo-siRNA is then loaded onto AGO2 in *Drosophila*. In addition to *Drosophila*, endo-siRNAs are found also in mice oocytes, embryonic stem cells, plants, and *C. elegans* [12]. In *C. elegans*, these endo-siRNAs can undergo amplification via RNA-dependent RNA polymerases to produce secondary endo-siRNAs. While targeting mechanisms are less well known, the sequences of endo-siRNAs match precisely to complementary sequences of its target, in contrast to seed pairing in miRNAs.

12.6 EXOGENOUS SILENCING RNAs (exo-siRNAs)

Exogenous silencing RNAs (exo-siRNAs) are exogenously sourced small RNAs which are approximately 21 nt in length. These are introduced to the cell via viral infection in a natural setting and via transfection or transduction of DNA sequences that are then endogenously transcribed to dsRNA sequences. An example of an exogenous source would be the Hepatitis C virus which is a double-stranded RNA virus. dsRNA sequences are recognized by the cell and digested by DICER2. The primary exo-siRNAs generated by DICER can be amplified by RNA-dependent RNA polymerases in *C. elegans*. Indeed, it has been shown that the exo- and endo-siRNA pathways in *C. elegans* may share some limiting components [13].

12.7 TRANSFER RNAs (tRNAs)

Transfer RNAs (tRNAs) are present in all organisms and vary in length from 73 to 95 nt. They are transcribed by RNA polymerase III and are functional based on their secondary and tertiary structures. tRNAs play a pivotal role in translation by bringing amino acids into the ribosome for protein synthesis. The number of tRNA genes from an organism ranges from 170 to 570 in the most commonly studied species, with humans having 497 [14].

12.8 SMALL NUCLEOLAR RNAs (snoRNAs)

Small nucleolar RNAs (snoRNAs) are 60–150 nt long nuclear RNAs whose role is to guide processing of ribosomal RNA molecules. It performs this task by associating with proteins in the small nucleolar ribosomal complex. Approximately 400 snoRNAs have been identified in humans to date [15]. An snoRNA contains 10–20 nucleotides antisense to its ribosomal RNA target and uses this to guide the modification of rRNAs that includes methylations and pseudouridylations. snoRNAs are typically located in introns of proteins involved in ribosome synthesis and are thus transcribed by RNA polymerase II, but can also be transcribed by their own promoter. snoRNAs have conserved structures: a C/D box with two short conserved sequence motifs and an H/ACA box. The H/ACA box of snoRNAs form two hairpins.

12.9 SMALL NUCLEAR RNAs (snRNAs)

Small nuclear RNAs (snRNAs) are approximately 150 nt long nuclear RNAs also known as U-RNAs due to their high uridine content. They are transcribed mainly by RNA polymerase II or by III (U6) and function to process heteronuclear RNA in small nuclear riboprotein complexes. Thus, snRNAs such as U1, U2, U4, U5, and U6 form part of the spliceosome that cuts out introns during mRNA processing. Higher organisms contain 5–30 snRNAs.

12.10 ENHANCER-DERIVED RNAs (eRNA)

Enhancer-derived RNAs (eRNA) are short RNAs obtained from sequencing products during nascent transcription that map to enhancer regions [16]. Nascent transcription products can be identified using immunoprecipitation and also global run-on sequencing (GRO-seq). The enhancers are located proximal to coding genes and the eRNAs can be transcribed in both sense and antisense orientation with regard to the proximal gene. Recent studies indicated that several thousand eRNAs may be produced in cell cultures after hormonal or electrophysiological stimulation [17]. Cell culture studies also indicate that eRNAs can either activate or repress transcription of their proximal genes likely through recruitment or interaction with factors governing the structure of chromatin loops that are known to form between enhancers and its proximal gene.

12.11 OTHER SMALL NONCODING RNAs

Other small noncoding RNAs include tRNA-derived RNA fragment (tRF), tRNA-derived small RNA (tsRNA), snoRNA-derived RNA (sdRNA), promoter-associated small RNA, and transcription start site-associated RNA (TSSa-RNA). These fragments may be 17–26 nt in length and though initially believed to be degradation products from tRNAs, snoRNAs, or transcription; nonetheless, their high abundance, distinct expression patterns, and phenotypic effects when knocked down by siRNAs suggest that they may be functional [18].

It is becoming rapidly apparent that many different classes of small RNAs exist. These can be distinguished by their size, biogenesis, processing, expression patterns, intracellular localization, and molecular function. When planning RNA-seq experiments, it is therefore imperative to first know the expected location and sizes one expects to find the small-RNA class. A rough estimation of the abundance can also be of practical use in determining the feasibility and extent of experimental effort and material required. It may also benefit in deciding the depth of coverage required. Using knowledge of the expected sequences can also help for annotating the sequence reads from an RNA-seq experiment. While a detailed view on each noncoding small-RNA class is beyond the scope of this book, some excellent reviews exist that can be helpful for the curious reader. Also, Table 12.3 that shows the comparative orthologs of miRNA pathway components should help the reader to navigate the various nomenclatures used by researchers in the human, *Drosophila*, and

TABLE 12.3 miRNA Pathway Component Genes and Their Orthologs in *Drosophila* and *C. elegans*

Component/Function	Human Gene Name	*Drosophila* Ortholog	*C. elegans* Ortholog
Drosha: cleaves primary miRNA transcripts in complex with DCGR8	DROSHA	drosha	*drsh-1*
DiGeorge syndrome critical region: Recognizes RNA substrate and cofactor for DROSHA	DGCR8	pasha	*pash-1*
Exportin-5: transfers precursor miRNAs from nucleus to cytoplasm	XPO5	Ranbp21	???
Dicer: RNAase III family protein cleaves dsRNA	DICER1	Dcr-1	*dcr-1*
Argonaute: binds dsRNA and processes precursor RNAs to guide and passenger strands, brings guide strand to target mRNA	AGO1-4	AGO1-2	*alg-1, alg-2*
PIWI proteins: binds piRNAs and helps to amplify and produce secondary piRNAs, is a specialized Argonaute protein	PIWI1-4	AUB, AGO3, PIWI	???
TRBP: *trans*-activation response (TAR) RNA-binding protein acts with Dicer to recognize dsRNA for cleavage, alters rate of cleavage and involved in editing	TRBP	loqs	*rde-4*
PACT: protein activator of PKR is a dsRNA-binding protein that assists possibly in strand selection and is in complex with Dicer, but does not assist cleavage activity	PACT	???	???
GW182: helps miRNA target and silence mRNAs in complex with translation initiation factors, adenosine deaminases, and decapping enzymers	GW182	Gawky	*ain-1, ain-2*

Symbol: ???, unknown.

C. elegans scientific research communities. The basic as well as detailed steps of miRNA biogenesis, maturation, expression, and RISC assembly can be found in an excellent review by Bartel [19]. The equivalent review on piRNA can be found in an article by Siomi et al. [20]. A short summary on endo-siRNA biogenesis can be found in the review by Kim et al. [21]. An excellent and up-to-date review on the miRNA and siRNA pathways, in general, and Argonautes, in particular, can be found in [22].

12.12 SEQUENCING METHODS FOR DISCOVERY OF SMALL NONCODING RNAs

The main principles for sequencing methods in small noncoding RNA sequencing follow very closely with those for RNA-seq as described in Chapter 1. The main difference arises from selecting and enriching the RNA pool to contain small RNAs. As with RNA-seq, RNA is isolated from a biological source, adaptors are added to the ends, a cDNA strand is synthesized, and the cDNA is amplified with PCR primers that contain indexes and sequencing primers to create the library. Size selection is performed both when the input RNA is purified or enriched for small RNAs and also after the cDNA strand is synthesized. In practice, size selection can also be performed after amplification of the library. New laboratory instruments that utilize lab on a chip electrophoresis technologies such as the Pippen-prep system can essentially be used to purify and size select at any stage of library preparation. Below are more detailed descriptions and workflows of the current major small-RNA-seq methods. They generally differ only at the beginning stage of small-RNA purification where a specific target population of small noncoding RNAs (e.g., Argonaute-bound small RNAs) is purified prior to library construction. The library construction then proceeds using a variety of commercial kits. Kits available to produce small-RNA libraries for sequencing have been optimized for the individual sequencing platforms. Below, we detail the sequencing approaches for the major small-RNA-seq methods.

12.12.1 microRNA-seq

miRNA-seq is similar to RNA-seq with the exception that input material is not polyadenylated and the size of the RNA aimed for sequencing is small. According to the table from the previous chapter, small RNAs can range from 21 nt to several hundreds depending upon the class one wishes to have sequenced in the library. Therefore, it is critical for library preparation that the input RNA is purified or selected. For a typical miRNA-seq experiment, where miRNAs are the targets to sequence, the miRNA is enriched or size selected from the pool of total RNAs. This can be easily accomplished using commercially available kits or gel purification schemes. There is even now a commercially available microfluidic chip system to aid in this step. The quality and quantity of small RNAs is an initial and critical step. Degraded RNAs will not only increase the number of artifacts in the library, but also make interpreting sequence information problematic. In practice, when we have a large number of samples to prepare, we use an RNA stability agent RNAlater (Ambion, Austin, TX/Life Technologies, Carlsbad, CA), so samples can be collected on one day and processed the next. We have noticed no effect of this reagent on the quality of our RNA samples. Indeed, it has helped facilitate transport and storage of our samples. For isolating small RNAs, we use a column purification method that enriches for small RNAs <200 nt. We have had success with mirVana kits (Ambion/Life Technologies), although others might work just as well. Once our small-RNA library has been synthesized, see protocol steps below, we keep them at –80°C or in dry ice before sending them to the sequence facility. Because of the comparisons we desire to make, it is normal practice to make many libraries at once in parallel. The library protocol below is based on sequencing on the Illumina platform. Please be aware that GAIIx and Hi-Seq 2000 instruments are compatible with different library adaptors, so be sure the sequence of your adaptors in your library protocol can be used with the instrument you plan to use. The library protocol generally takes 3 days. The general workflow of cDNA library preparation is shown in Figure 12.2. A detailed example, without use of kits, on the workflow and adaptor sequences can be found from Juhila et al. [23].

miRNA-seq library workflow

1. (Optional) Place your sample in 10 volumes of RNAlater™ in a microfuge tube. Leave overnight at 4°C. After overnight storage, samples can be stored frozen in –20°C for months or –80°C indefinitely.

2. (Optional) Remove as much RNAlater from your samples as possible by centrifugation and pipetting off residual RNAlater™.

3. Isolate small RNA <200 nt enriched RNAs using mirVana™ Kit according to the manufacturer's instructions. In the final elution step, use 60 μL per column instead of 100 μL as recommended. We have also used miRNeasy Kits (Qiagen) with success. Both kits are

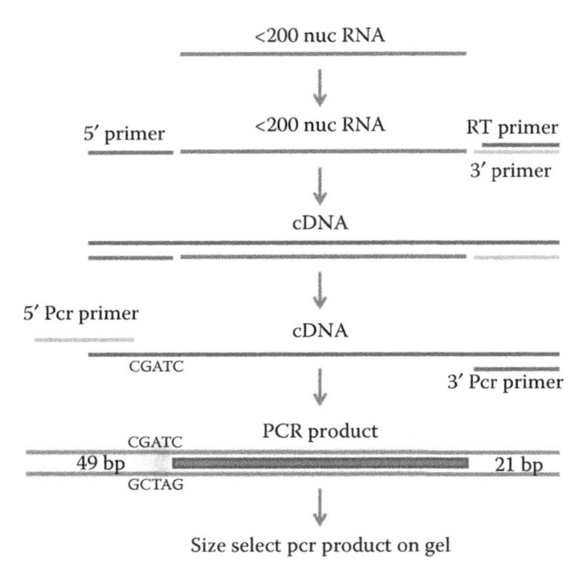

FIGURE 12.2 Scheme for small RNA library preparation. Small RNAs are first enriched using biochemical methods. A 3′ adaptor and 5′ adaptor are ligated to the RNA molecules. Reverse transcription is performed with a primer complementary to the 3′ adaptor to produce a cDNA of the RNA. PCR is then performed on the cDNA using 5′ and 3′ PCR primers based on the ligated adaptor sequences. The 5′ primer has an overhang to the cDNA sequence where library indexes can be used for multiplexing.

easy to use, handy, and provide sufficient quality of small RNA for library preparation.

4. Use a 2 μL aliquot to measure the RNA concentration and recovery with Nanodrop (Thermo Scientific) or Qubit®fluorometer (Life Technologies) methods. Nanodrop is more convenient, while Qubit tends to be more accurate, especially with low yields. A typical yield is a total of 2–4 μg of enriched small RNAs in 50 μL volume from an original tissue pellet of 0.5 mL.

5. Dry down the small-RNA pellet to approximately 5 μL in a speedvac without heating.

6. Prepare the small-RNA sample library using NEBNext Small RNA Sample Prep Set 1 Kit (New England Biolabs) or TruSeq® small-RNA sample preparation kit (Illumina), according to the manufacturer's instructions. In the PCR amplification step of the library, use 12

cycles. The sample library kit must be compatible with the sequencing platform you plan to use. The kit's technical manuals mention this. New England Biolabs, for example, has different small-RNA sample prep kits that are separately compatible for Illumina, Solid, Ion Torrent, and 454 platforms. Each platform has their own kits and these, you can be certain, are compatible.

7. In the size selection and gel purification of the amplified cDNA library, cut the 80–110 nt band. Avoid the major band at 70–75 bp as this contains the adaptor dimer. The mature miRNAs are at 91–93 nt due to ligation of adaptors, sequence primers, and indices. Cut larger size bands if interested in pre-miRNAs.

8. Elute the amplified cDNA library from the acrylamide gel as directed in the kit manufacturer's instructions, resuspending the pellet in 10 μL of TE buffer.

9. Use 1 μL of the cDNA library to view the size on a Bioanalyzer (Agilent Technologies). See Figure 12.3 for an example of an miRNA library analyzed for size.

10. If the size of the library is as expected, continue by generating clusters with platform-specific cluster kit (TruSeq SR Cluster Kit, Illumina) and sequencing by synthesis kit (TruSeq SBS Kit, Illumina). This last step of cluster generation and sequencing is typically performed by the core unit.

12.12.2 CLIP-seq

Cross-linked immunoprecipitation sequencing (CLIP-seq) is a variation of RNA-seq that differs principally in the source of RNA for the library. Early studies to identify miRNA:mRNA molecules used immunoprecipitation as a tool to enrich interacting small-RNA molecules in complex with proteins. These studies used antibodies against myc-tagged Argonaute proteins to precipitate bound RNA molecules and then proceeded to identify mRNA target molecules [24] or analyze them using microarrays [25]. The purpose was to isolate for sequencing both mature miRNAs and their targets. A further improvement on this method utilized first crosslinking of RNA molecules with proteins in the immediate physical vicinity via UV irradiation, and, secondly, RNA-seq instead of cloning or microarray analysis to identify RNA molecules and was given

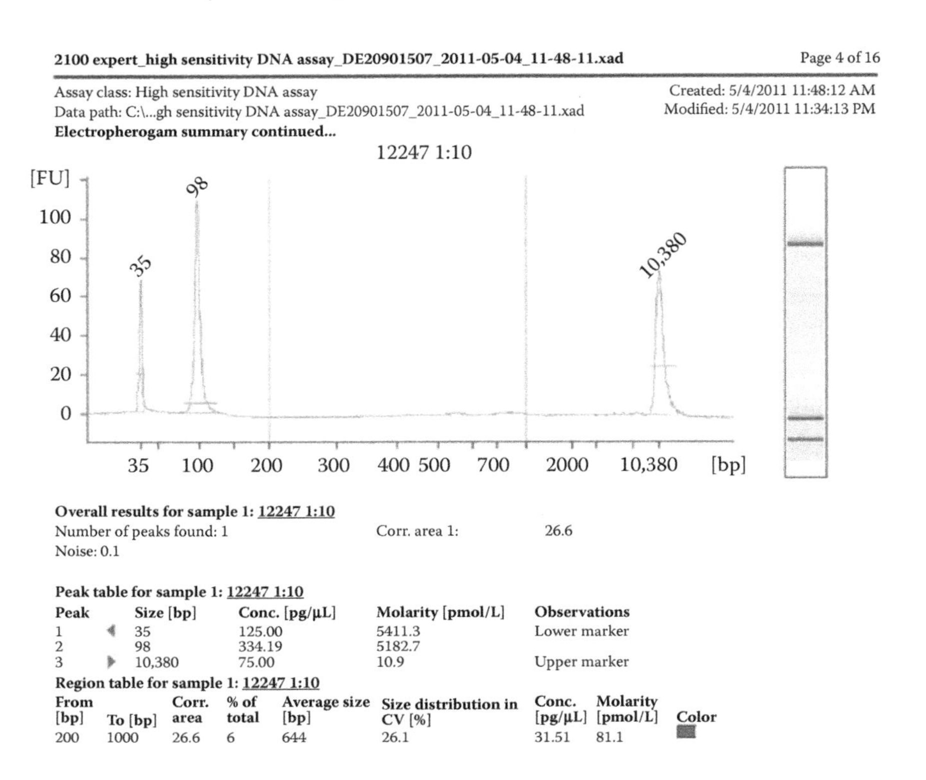

Peak table for sample 1: 12247 1:10

Peak	Size [bp]	Conc. [pg/µL]	Molarity [pmol/L]	Observations
1	35	125.00	5411.3	Lower marker
2	98	334.19	5182.7	
3	10,380	75.00	10.9	Upper marker

Region table for sample 1: 12247 1:10

From [bp]	To [bp]	Corr. area	% of total	Average size [bp]	Size distribution in CV [%]	Conc. [pg/µL]	Molarity [pmol/L]	Color
200	1000	26.6	6	644	26.1	31.51	81.1	

FIGURE 12.3 Electropherogram of an miRNA library run on Agilent Bioanalyzer. Size markers are at 35 and 10,380 bp, while the library shows a peak at 98 bp. In the 98 bp peak, the small RNAs are 21–23 nt, while the rest of the dsDNA consists of the library adaptors.

the name high-throughput sequencing of RNA isolated by crosslinking immunoprecipitation (HITS-CLIP, now usually shorted to simply CLIP-seq) [26]. While the initial study immunoprecipitated the neuron-specific splicing factor Nova, a later study crosslinked and immunoprecipitated Argonaute protein (Ago) from mouse brain in a complex with miRNA and mRNA [27]. The results yielded mRNA sequence from 829 transcripts. A similar study was performed in *C. elegans*, using wild-type and Argonaute (*alg-1*) null mutants as a control, and through this method, 3093 genes specific to the wild-type were identified [28]. RNA-binding protein immunoprecipitation sequencing (RIP-seq) is also sometimes used as a term to describe CLIP-seq when the epitope is the RNA-binding protein [29]. CLIP-seq has also been applied to identify RNAs in the pri-miRNA processor complex via immunoprecipitation of DGCR8 [30]. Interestingly, they find pri-miRNAs as well as mRNAs and snoRNAs, suggesting additional functions for DGCR8. As the recovery of small RNAs is miniscule from immunoprecipitated samples, the use of negative controls becomes very important. For this reason, multiple replicates, as well as background controls, have extensively been used in CLIP-seq protocols. An excellent step-by-step protocol for CLIP-seq can be found from the following publication [31].

Photoactivatable ribonucleoside-enhanced CLIP (PAR-CLIP Seq) is a variation on CLIP-seq whose purpose is to map the precise interaction sites between the protein and RNA at the nucleotide level. PAR-CLIP uses modified ribonucleotides such as 4-thiouridine that are incorporated into the RNA molecule during biosynthesis. The modified photoreactive ribonucleotides are then crosslinked to the RNA-binding protein via UV light in a reaction that is much more efficient than the nonmodified nucleotide [32]. Another advantage of this technique is that more precise mapping of the site of RNA::protein interaction can be obtained. After UV crosslinking, immunoprecipitation and next-generation sequencing subsequently take place in the same manner as CLIP-seq. A drawback of this technique is that it requires the preincubation of cells/samples with nucleoside analogs, thus limiting the range of samples applicable to this technique. Moreover, some ribonucleoside analogs may be toxic and thus bias results where comparisons are made between samples.

Individual-Nucleotide Resolution UV Crosslinking and Immunoprecipitation (iCLIPSeq) is another variation on CLIP-seq. Early CLIP-seq studies showed that some immunoprecipitated RNAs contained missing sequences when mapped to their respective genome. This was presumed due to reverse transcriptase skipping over RNA sequences where crosslinked protein remained bound. To improve resolution, iCLIPSeq was created to map the RNA:protein interaction to the single-nucleotide level [33]. The method relies on the observation that during library construction, in the step where the immunoprecipitated and isolated RNA is being reverse-transcribed to cDNA, the reverse transcriptase enzyme either skips over or stops at the exact point where the RNA and protein are crosslinked in the few molecules where proteinase K treatment did not completely dissociate the protein from the RNA. In these instances, the resulting truncated cDNA is circularized, linearized, amplified, and then sequenced [34]. The resulting sequence provides not only identified immunoprecipitated RNAs, but also protein interaction sites at the individual nucleotide level.

12.12.3 Degradome-seq

Motivated by the question of where in the sequence and how mRNA molecules are cleaved by endoribonucleases, a set of methods now better known as Degradome-seq have been created [35,36]. The most practical application for this method has been to map, match, and identify miRNAs and the cleavage product of their target mRNAs. This method relies on the principle biochemical difference that mature eukaryotic mRNAs are capped at the 5′ end with 7-methylguanasine while endoribonuclease cleaved mRNAs have a 5′-monophosphate. In the method, small-RNA libraries are made from cellular RNAs containing a 5′-monophosphates that exclude noncleaved mRNAs since 7-methylguanasine nucleotides cannot ligate the adaptors during library construction. Libraries are then subjected to RNA-seq. Reads are then mapped to the genome and the 5′ end where it maps becomes a potential site of miRNA-mediated cleavage. If the cleavage was mediated by dicer, aRNAse III class endonuclease, the beginning of the miRNA should be precisely 10 nucleotides from the 5′ end of the miRNA. Thus, the miRNA sequence can be inferred from the cleavage site. Cross-checking the sequences surrounding the cleavage site with various miRNA databases can verify the miRNA sequence and identity. Other bioinformatic methods such as calculating RNA secondary structure and minimum free energy of hairpins in the vicinity of the cleave site can also be used to identify novel miRNAs. This combined library construction, sequencing, and bioinformatics analysis set of methods has also been termed PARE for parallel analysis of RNA ends. The method was first described in plants, but has now been used in many metazoan species. A caveat to the method is that 5′ monophosphates can be generated from other cellular RNAse III class molecules apart from dicer and is therefore more representative of endogenous RNA degradation activity. Therefore, the term for this method, degradeome-seq, seems appropriate. A notable computational and technical challenge for this method is the mapping and identification of miRNAs. A bioinformatic pipeline, CleaveLand, has been produced and can detect cleaved miRNA targets from degradome-seq projects [35]. CleaveLand can be downloaded from http://axtell-lab-psu.weebly.com/cleaveland.html.

12.12.4 Global Run-On Sequencing (GRO-seq)

Global run-on sequencing (GRO-seq) describes a method to identify nascent, newly synthesized transcripts as they are being elongated. It identifies, at the moment of assay, transcripts that are actually being synthesized, in contrast to steady-state levels of transcripts in a biological sample. It is a method that exploits the occurrence of transcriptional pausing, where the RNA polymerase has started to transcribe or is engaged in transcriptional activity, but the transcription complex may be waiting for additional factors and therefore has paused. Pausing or transcriptional interference can also be obtained using toxins such as actinomycin D. To obtain RNA fragments for sequence in this context, nuclear run-on is performed where 5-bromouridine 5′ phosphate is added to culture and incorporated into nascent RNA being produced in the cell [37]. It is most typically used to identify actively transcribed genes but more recently has been used to identify eRNAs.

12.13 SUMMARY

The common theme of small-RNA seq approaches is the isolation of small RNAs and their subsequent sequencing on various platforms. As scientists continue to find novel classes of small RNAs, even very recently such as eRNAs, and each organism seems to have its own unique miRNAome, these methods will continue to be developed. Indeed, many methods now include fractionation steps that are intended to enrich RNAs for both small RNA and mRNA library synthesis and generation. Utilizing the wealth of technology currently available, it appears that many new and exciting discoveries remain to be made.

REFERENCES

1. Lee R.C., Feinbaum R.L., and Ambros V. The *C. elegans* heterochronic gene lin-4 encodes small RNAs with antisense complementarity to lin-14. *Cell* 75(5):843–854, 1993.
2. Pasquinelli A.E., Reinhart B.J., Slack F. et al. Conservation of the sequence and temporal expression of let-7 heterochronic regulatory RNA. *Nature* 408(6808):86–89, 2000.
3. Lagos-Quintana M., Rauhut R., Lendeckel W. et al. Identification of novel genes coding for small expressed RNAs. *Science* 294(5543):853–858, 2001.
4. Lau N.C., Lim L.P., Weinstein E.G. et al. An abundant class of tiny RNAs with probable regulatory roles in *Caenorhabditis elegans*. *Science* 294(5543):858–862, 2001.
5. Lee R.C. and Ambros V. An extensive class of small RNAs in *Caenorhabditis elegans*. *Science* 294(5543):862–864, 2001.
6. Djuranovic S., Nahvi A., and Green R. A parsimonious model for gene regulation by miRNAs. *Science* 331(6017):550–553, 2011.

7. Neilsen C.T., Goodall G.J., and Bracken C.P. IsomiRs—the overlooked repertoire in the dynamic microRNAome. *Trends Genet* 28(11):544–549, 2012.
8. Westholm J.O. and Lai E.C. Mirtrons: MicroRNA biogenesis via splicing. *Biochimie* 93(11):1897–1904, 2011.
9. Shi W., Hendrix D., Levine M. et al. A distinct class of small RNAs arises from pre-miRNA-proximal regions in a simple chordate. *Nat Struct Mol Biol* 16(2):183–189, 2009.
10. Langenberger D., Bermudez-Santana C., Hertel J. et al. Evidence for human microRNA-offset RNAs in small RNA sequencing data. *Bioinformatics* 25(18):2298–2301, 2009.
11. Bortoluzzi S., Biasiolo M., and Bisognin A. MicroRNA-offset RNAs (moRNAs): By-product spectators or functional players? *Trends Mol Med* 17(9):473–474, 2011.
12. Asikainen S., Heikkinen L., Wong G. et al. Functional characterization of endogenous siRNA target genes in *Caenorhabditis elegans*. *BMC Genomics* 9:270, 2008.
13. Duchaine T.F., Wohlschlegel J.A., and Kennedy S. Functional proteomics reveals the biochemical niche of *C. elegans* DCR-1 in multiple small-RNA-mediated pathways. *Cell* 124(2):343–354, 2006.
14. Goodenbour J.M. and Pan T. Diversity of tRNA genes in eukaryotes. *Nucleic Acids Res* 34(21):6137–6146, 2006.
15. Lestrade L. and Weber M.J. snoRNA-LBME-db, a comprehensive database of human H/ACA and C/D box snoRNAs. *Nucleic Acids Res* 34(Database issue):D158–D162, 2006.
16. Redmond A.M. and Carroll J.S. Enhancer-derived RNAs: "spicing up" transcription programs. *EMBO J* 32(15):2096–2098, 2013.
17. Kim T.K., Hemberg M., Gray J.M. et al. Widespread transcription at neuronal activity-regulated enhancers. *Nature* 465(7295):182–187, 2010.
18. Aalto A.P. and Pasquinelli A.E. Small non-coding RNAs mount a silent revolution in gene expression. *Current Opinion Cell Biol* 24(2):333–340, 2012.
19. Bartel D. MicroRNAs: Genomics, biogenesis, mechanism, and function. *Cell* 116(2):281–297, 2004.
20. Siomi M.C., Sato K., Pezic D. et al. PIWI-interacting small RNAs: The vanguard of genome defence. *Nat Rev Mol Cell Biol* 12(4):246–258, 2011.
21. Kim V.N., Han J., and Siomi M.C. Biogenesis of small RNAs in animals. *Nat Rev Mol Cell Biol* 10(2):126–139, 2009.
22. Meister G. Argonaute proteins: Functional insights and emerging roles. *Nat Rev Genet* 14(7):447–459, 2013.
23. Juhila J., Sipilä T., Icay K. et al. MicroRNA expression profiling reveals miRNA families regulating specific biological pathways in mouse frontal cortex and hippocampus. *PLoS ONE* 6(6):e21495, 2011.
24. Karginov F.V., Conaco C., Xuan Z. et al. A biochemical approach to identifying microRNA targets. *Proc Natl Acad Sci USA* 104(49):19291–19296, 2007.
25. Easow G., Teleman A.A., and Cohen S.M. Isolation of microRNA targets by miRNP immunopurification. *RNA* 13(8):1198–1204, 2007.

26. Licatalosi D.D., Mele A., Fak J.J. et al. HITS-CLIP yields genome-wide insights into brain alternative RNA processing. *Nature* 456(7221):464–469, 2008.
27. Chi S.W., Zang J.B., Mele A. et al. Argonaute HITS-CLIP decodes microRNA–mRNA interaction maps. *Nature* 460(7254):479–486, 2009.
28. Zisoulis D.G., Lovci M.T., Wilbert M.L. et al. Comprehensive discovery of endogenous Argonaute binding sites in *Caenorhabditis elegans*. *Nat Struct Mol Biol* 17(2):173–179, 2010.
29. Zhao J., Ohsumi T.K., Kung J.T. et al. Genome-wide identification of polycomb-associated RNAs by RIP-seq. *Mol Cell* 40(6):939–953, 2010.
30. Macias S., Plass M., Stajuda A. et al. DGCR8 HITS-CLIP reveals novel functions for the microprocessor. *Nat Struct Mol Biol* 19(8):760–766, 2012.
31. Murigneux V., Saulière J., Roest Crollius H. et al. Transcriptome-wide identification of RNA binding sites by CLIP-seq. *Methods* 63(1):32–40, 2013.
32. Hafner M., Lianoglou S., Tuschl T. et al. Genome-wide identification of miRNA targets by PAR-CLIP. *Methods* 58(2):94–105, 2012.
33. König J., Zarnack K., Rot G. et al. iCLIP reveals the function of hnRNP particles in splicing at individual nucleotide resolution. *Nat Struct Mol Biol* 17(7):909–915, 2010.
34. Sugimoto Y., König J., Hussain S. et al. Analysis of CLIP and iCLIP methods for nucleotide-resolution studies of protein–RNA interactions. *Genome Biol* 13(8):R67, 2012.
35. Addo-Quaye C., Eshoo T.W., Bartel D.P. et al. Endogenous siRNA and miRNA targets identified by sequencing of the *Arabidopsis* degradome. *Current Biol* 18(10):758–762, 2008.
36. German M.A., Pillay M., Jeong D.H. et al. Global identification of microRNA-target RNA pairs by parallel analysis of RNA ends. *Nat Biotechnol* 26(8):941–946, 2008.
37. Core L.J., Waterfall J.J., and Lis J.T. Nascent RNA sequencing reveals widespread pausing and divergent initiation at human promoters. *Science* 322(5909):1845–1848, 2008.

CHAPTER 13

Computational Analysis of Small Noncoding RNA Sequencing Data

13.1 INTRODUCTION

After small RNAs have been isolated from your samples, libraries constructed, and sequence data obtained, you will normally be provided with a large file or files containing your sequences in FASTA format. Now the real fun begins! There are many tools available to analyze the data depending upon the aims of the experiment. The tools differ in many dimensions including ease of use, algorithms to detect small RNA species, dependencies on other data files and annotations, statistical methods, add-ons, and input and output formats, just to name a few factors. We take you through a step-by-step practical approach with *miRdeep2* since this works well in our hands, has been used and cited by other researchers in the field, and contains inclusive tools for mapping reads, folding and visualization of potentially newly discovered miRNAs, and has an intuitive easy to read output. It does, however, require working knowledge of the command line. As a second tool, we demonstrate the use of *miRanalyzer* which is a web-based tool and can be run on any web-enabled browser, is easy to use, and provides tools for differential analysis for which many studies would be interested in performing. These are not the only tools for analysis of small RNA-seq data. The two examples we provide are intended to provide a flavor of practical solutions that are relatively simple or extremely simple to implement. For readers interested in other available tools, a couple of recent reviews also detail the differences and evaluate different strengths and weaknesses of tools shown here and others currently available [1,2]. In the second part of the chapter, we look at practical approaches to solve downstream analysis once the miRNAs have been identified and quantitated from your precious samples. These analyses include locating the mRNA targets of the miRNAs. Finally, we point out some important sources of general information regarding small RNAs including *miRBase* and *RFAM*.

13.2 DISCOVERY OF SMALL RNAs—miRDeep2

miRDeep2 is a comprehensive tool that allows you to input RNA-seq data and obtain as output counts of known miRNAs, discovery of novel miRNAs, and differences in miRNAs between data sets [3]. The software runs in Linux. The modules within the perl wrapper include a tool for mapping reads (Bowtie), and tools for folding and visualization of miRNA precursors (Randfold, and Viennarna). The command line interface and graphical output make this a very useful tool.

The basic method to run miRDeep2 follows:

1. Install Linux on your workstation. For specific instructions on how to do this, see CHAPTER 2.6.
2. From your Linux installation, download miRDeep2 (https://www.mdc-berlin.de/8551903/en/research/research_teams/systems_biology_of_gene_regulatory_elements/projects/miRDeep).
3. Download the gff annotation file for the species you want to analyze.
4. Download the FASTA file of known miRNAs for the species you want to analyze.
5. Set up your running environment, for example, place all files you need in a "mirDeep2 working directory" and configure the scripts as needed.
6. Run miRDeep2.

13.2.1 GFF files

GFF files—GFF stands for Generic Feature Format and contains the genomic features of a sequence. The features are stored in a plain text file format with nine columns, with each column representing a distinct feature of a sequence. The columns are tab-delimited. The first line of the file contains a text header ##gff-version 3. The following lines can also contain descriptor text such as the description of the file, origin, version number, notes, references, and descriptions as long as the line begins with the # symbol. After the descriptor lines, the genomic features can be seen in each of the nine columns. In the example shown in Table 13.1, the nine lines contain the following genomic feature information.

Column 1. Landmark ID, in this case it is I to denote chromosome I.

Column 2. Source that generated this feature. It is left empty in the example and a period is placed when there is no source listed in the field.

Column 3. Type of sequence, in this case miRNA primary transcript, or as seen after this entry miRNA.

Column 4. Start of the sequence according to the coordinate system. In the example, 1738637 is shown meaning the sequence starts at chromosome I, nucleotide 1738637.

Column 5. End of the sequence according to the coordinate system. In the example, 1738735 is shown meaning the sequences ends at chromosome I, nucleotide 1738735.

Column 6. Score, in this case left empty with an ..

Column 7. Strand, in this case +, indicating the positive strand.

Column 8. Phase, in this case, left empty with an.

Column 9. Attributes of the sequence. The system used is a tag = value, separated by a semicolon. Each tag is an attribute and can be given a value. Multiple tags can be used. In our example, the first tag is the ID which is MI0000021-1, the second tag is NameM which is cel-mir-50.

GFF files come in different versions and we use version 3. Beware that although the versions are similar, they are not always compatible. GFF files we used were downloaded from miRBase, but can be downloaded from many different sources such as different genome databases. Precise specifications can be found at www.sequenceontology.org/gff3.shtml.

TABLE 13.1 A GFF File of *C. elegans* miRNAs

```
##gff-version 3
##date
 2012-7-23
#
# Chromosomal coordinates of C. elegans microRNAs
# microRNAs                                miRBase v19
#                   genome-build-id WBcel215
#
# Hairpin precursor sequences have type "miRNA_primary_transcript"
# Note, these sequences do not represent the full primary transcript
# rather a predicted stem-loop portion that includes the precursor
# miRNA. Mature sequences have type "miRNA."
#
```

I	.	miRNA_primary_transcript	17,38,637	17,38,735	.	+	.	ID = MI0000021_1;Name = cel-mir-50
I	.	miRNA	17,38,652	17,38,675	.	+	.	ID = MIMAT0000021_1;Name = cel-miR-50-5p
I	.	miRNA	17,38,694	17,38,715	.	+	.	ID = MIMAT0020310_1;Name = cel-miR-50-3p
I	.	miRNA_primary_transcript	28,88,450	28,88,559	.	-	.	ID = MI0019067_1;Name = cel-mir-5546
I	.	miRNA	28,88,514	28,88,536	.	-	.	ID = MIMAT0022183_1;Name = cel-miR-5546-5p
I	.	miRNA	28,88,472	28,88,493	.	-	.	ID = MIMAT0022184_1;Name = cel-miR-5546-3p
I	.	miRNA_primary_transcript	29,21,188	29,21,292	.	+	.	ID = MI0017717_1;Name = cel-mir-4931
I	.	miRNA	29,21,258	29,21,277	.	+	.	ID = MIMAT0020137_1;Name = cel-miR-4931

The hash sign (#) indicates comment lines. The 9 columns of data follow the comments. The table has been abbreviated after the 8th entry.

13.2.2 FASTA Files of Known miRNAs

The FASTA file format is a text format for DNA, RNA, or protein sequences. The first line of a FASTA formatting sequence contains a greater than symbol ">" followed by text describing the sequence. The rest of the lines contain the sequence itself. The next sequence begins when the next ">" symbol occurs. FASTA files for known miRNAs for each species can be downloaded directly from miRBase. For example, the FASTA file for *Caenorhabditis elegans* miRNAs is shown in Table 13.2.

13.2.3 Setting up the Run Environment

1. Make sure GCC tool chain has been installed. In Ubuntu, install package build-essential:

   ```
   $ sudo apt-get install build-essential
   ```

2. Get Bowtie (bowtie-bio.sourceforge.net). Unzip the package to/usr/local/share and make symbolic links to/usr/local/bin to point to the unzipped executable files:

   ```
   $ sudo unzip name_of_the_bowtie_package.zip -d/usr/local/share
   ```

   ```
   $ ln -s/usr/local/share/name_of_the_bowtie_directory/bowtie*/usr/local/bin
   ```

3. Get Vienna RNA package (http://www.tbi.univie.ac.at/ ~ivo/RNA/). To compile and install, type

   ```
   $./configure
   ```

   ```
   $ make
   ```

   ```
   $ sudo make install
   ```

4. Install SQUID. In Ubuntu, type

   ```
   $ sudo apt-get install biosquid
   ```

5. Get the version 2 (C version) of Randfold

   (http://bioinformatics.psb.ugent.be/software/details/Randfold). First, modify the Makefile and add -I/usr/include/biosquid to the INCLUDE line (the line should now read INCLUDE = -I. -I/usr/include/biosquid). To compile and install, type

TABLE 13.2 FASTA File of *C. elegans* miRNAs

>cel-let-7 MI0000001 *C. elegans* let-7 stem-loop
UACACUGUGGAUCCGGUGAGGUAGUAGGUUGUAUAGUUUGGAAUAUUACCACCGGUGAAC
UAUGCAAUUUUCUACCUUACCGGAGACAGAACUCUUCGA
>cel-lin-4 MI0000002 *C. elegans* lin-4 stem-loop
AUGCUUCCGGCCUGUUCCCUGAGACCUCAAGUGUGAGUGUACUAUUGAUGCUUCACACCU
GGGCUCUCCGGGUACCAGGACGGUUUGAGCAGAU
>cel-mir-1 MI0000003 *C. elegans* miR-1 stem-loop
AAAGUGACCGUACCGAGCUGCAUACUUCCUUACAUGCCCAUACUAUAUCAUAAAUGGAUA
UGGAAUGUAAAGAAGUAUGUAGAACGGGGUGGUAGU
>cel-mir-2 MI0000004 *C. elegans* miR-2 stem-loop
UAAACAGUAUACAGAAAGCCAUCAAAGCGGUGGUUGAUGUGUUGCAAAUUAUGACUUUCA
UAUCACAGCCAGCUUUGAUGUGCUGCCUGUUGCACUGU
>cel-mir-34 MI0000005 *C. elegans* miR-34 stem-loop
CGGACAAUGCUCGAGAGGCAGUGUGGUUAGCUGGUUGCAUAUUUCCUUGACAACGGCUAC
CUUCACUGCCACCCCGAACAUGUCGUCCAUCUUUGAA
>cel-mir-35 MI0000006 *C. elegans* miR-35 stem-loop

UCUCGGAUCAGAUCGAGCCAUUGCUGGUUUCUUCCACAGUGGUACUUUCCAUUAGAACUA

UCACCGGGUGGAAACUAGCAGUGGCUCGAUCUUUUCC

>cel-mir-36 MI0000007 *C. elegans* miR-36 stem-loop

CACCGCUGUCGGGGAACCGCGCCAAUUUUCGCUUCAGUGCUAGACCAUCCAAAGUGUCUA

UCACCGGGUGAAAAUUCGCAUGGGUCCCCGACGCGGA

>cel-mir-37 MI0000008 *C. elegans* miR-37 stem-loop

UUCUAGAAACCCUUGGACCAGUGUGGGUGUCCGUUGCGGUGCUACAUUCUCUAAUCUGUA

UCACCGGGUGAACACUUGCAGUGGUCCUCGUGGUUUCU

>cel-mir-38 MI0000009 *C. elegans* miR-38 stem-loop

GUGAGCCAGGUCCUGUUCCGGUUUUUUCCGUGGUGAUAACGCAUCCAAAAGUCUCUAUCA

CCGGGAGAAAAACUGGAGUAGGACCUGUGACUCAU

>cel-mir-39 MI0000010 *C. elegans* miR-39 stem-loop

UAUACCGAGAGCCCAGCUGAUUUCGUCUUGGUAAUAAGCUCGUCAUUGAGAUUAUCACCG

GGUGUAAAUCAGCUUGGCUCUGGUGUC

$ make

$ sudocprandfold/usr/local/bin

6. Install PDF::API2 package. In Ubuntu, type

   $ sudo apt-get install libpdf-api2-perl

7. Get miRDeep2 (www.mdc-berlin.de/8551903/en/research/research_teams/systems_biology_of_gene_regulatory_elements/projects/miRDeep). Unzip and run.

13.2.4 Running miRDeep2

mapper.pl ctrl_trimmed.fasta -c -p/home/wong/rna_seq/genomes/ws220/genome -t ctrl_trimmed_mapped.arf -o 4 -n -s ctrl_trimmed_processed.fa -v -m

miRDeep2.pl ctrl_trimmed_processed.fa/home/wong/rna_seq/genomes/ws220/genome.fa ctrl_trimmed_mapped.arfmature.fa none hairpin.fa

miRDeep2 hints:

1. Move programs and required files into public/common directory so that it will always be in the path.

2. Create a new subdirectory to run miRdeep2 and to have your output files deposited there.

3. Move needed input files to the new miRdeep2 subdirectory.

13.2.4.1 miRDeep2 Output

One of the nice features of miRDeep2 is that the output files contain an html page that collects all of the output and allows you to read it in a familiar webpage format. While you can still look at the individual results, the use of the webpage links all of the output including graphics that are in separate pdf files. The output files are placed in the working directory where the program is executed unless otherwise specified. From the working directory, you click the html icon with the date and time stamp. In this example, the file is /mirdeepworking/expression_02_04_2013_t_15_15_09.html (Figure 13.1). The first set of outputs detail the parameters used including the version of miRDeep2, the program call, the name of the file where the reads were taken, the location and name of the genome file, the name of the map file, the name of the reference mature miRNA file, and any other mature miRNAs. The next section of

Parameters used

miRDeep2 version	2.0.0.5
Program call	/home/wong/mirdeep2/miRDeep2.pl ctrl_processed.fa /home/wong/rna_seq/genomes/ws220/genome.fa ctrl_mapped.arf mature.fa none hairpin.fa
Reads	ctrl_processed.fa
Genome	/home/wong/rna_seq/genomes/ws220/genome.fa
Mappings	ctrl_mapped.arf
Reference mature miRNAs	mature.fa
Other mature miRNAs	none

Survey of miRDeep2 performance for score cut-offs -10 to 10

	novel miRNAs			known miRBase miRNAs				
miRDeep2 score	predicted by miRDeep2	estimated false positives	estimated true positives	in species	in data	detected by miRDeep2	estimated signal-to-noise	excision gearing
10	3	2 ± 1	1 ± 1 (46 ± 37%)	112	112	2 (2%)	2.3	2
9	3	2 ± 1	1 ± 1 (45 ± 37%)	112	112	2 (2%)	2.2	2
8	3	2 ± 1	1 ± 1 (42 ± 37%)	112	112	2 (2%)	2.1	2
7	3	2 ± 2	1 ± 1 (41 ± 37%)	112	112	2 (2%)	2	2
6	3	2 ± 2	1 ± 1 (39 ± 36%)	112	112	2 (2%)	1.9	2
5	3	2 ± 2	1 ± 1 (38 ± 35%)	112	112	2 (2%)	1.8	2
4	3	2 ± 2	1 ± 1 (36 ± 35%)	112	112	2 (2%)	1.7	2
3	5	2 ± 2	3 ± 2 (53 ± 30%)	112	112	2 (2%)	2.4	2
2	22	4 ± 2	18 ± 2 (83 ± 9%)	112	112	62 (55%)	14.6	2
1	54	8 ± 3	46 ± 3 (85 ± 5%)	112	112	90 (80%)	12.3	2
0	69	40 ± 6	29 ± 6 (42 ± 9%)	112	112	93 (83%)	2.9	2
-1	79	64 ± 8	15 ± 7 (19 ± 9%)	112	112	93 (83%)	2	2
-2	99	91 ± 8	9 ± 7 (9 ± 7%)	112	112	97 (87%)	1.6	2
-3	124	123 ± 9	4 ± 6 (3 ± 5%)	112	112	97 (87%)	1.4	2
-4	137	166 ± 10	0 ± 0 (0 ± 0%)	112	112	98 (88%)	1.1	2
-5	150	223 ± 11	0 ± 0 (0 ± 0%)	112	112	100 (89%)	0.9	2
-6	166	271 ± 12	0 ± 0 (0 ± 0%)	112	112	100 (89%)	0.8	2
-7	188	319 ± 12	0 ± 0 (0 ± 0%)	112	112	100 (89%)	0.8	2
-8	220	369 ± 13	0 ± 0 (0 ± 0%)	112	112	100 (89%)	0.7	2
-9	257	419 ± 15	0 ± 0 (0 ± 0%)	112	112	100 (89%)	0.7	2
-10	281	474 ± 16	0 ± 0 (0 ± 0%)	112	112	100 (89%)	0.7	2

FIGURE 13.1 Output file from miRDeep2 showing the performance scores.

the output provides a survey of miRDeep2 performance with the number of identified novel-predicted miRNAs and known miRNAs. The third section provides a list of novel miRNAs predicted by miRDeep2 that includes the provisional id, miRDeep2 score, read counts in mature, loop, and star regions, randfold significance, links to external databases, link to NCBI blastn results, consensus mature sequence, consensus star sequence, consensus precursor sequence, and precursor coordinates by chromosome and location (Figure 13.2). The fourth section provides the list of mature miRBasemiRNAs with a tag id, miRDeep2 score, estimated probability that miRNA is a true positive, statement on agreement with miRBase mature sequence, total read counts of mature, loop and star forms, significance of randfold, the name of the mature miRBasemiRNA, link to NCBI blastn search, and the consensus mature, star, and hairpin sequences.

Clicking the provisional ID of a novel miRNA will give you an RNAfold-generated graphic containing the actual in silico-folded hairpin, with the number of reads for each part of the hairpin, score for minimum free energy, score for randfold, and score for conserved seed sequence. In addition, an alignment is generated for the reads mapping to the hairpin, including closely mapping reads which allows one to view any iso-mirs. An example is shown in Figure 13.3.

novel miRNAs predicted by miRDeep2

provisional id	miRDeep2 score	estimated probability that the miRNA candidate is a true positive	rfam alert	total read count	mature read count	loop read count	star read count	significant randfold p-value
X_16668	4.2e+2	0.46 ± 0.37		826	753	0	73	yes
IV_7271	2.1e+1	0.46 ± 0.37		37	35	1	1	yes
IV_12490	2.1e+1	0.46 ± 0.37		36	35	0	1	yes
X_17044	3.2	0.53 ± 0.30		818	748	11	59	yes
X_16689	3.2	0.53 ± 0.30		804	748	0	56	yes
IV_14843	2.9	0.83 ± 0.09		121	121	0	0	yes
IV_5013	2.8	0.83 ± 0.09		594	573	0	21	yes
II_2882	2.7	0.83 ± 0.09		42	41	0	1	yes
X_16990	2.4	0.83 ± 0.09		70	70	0	0	yes
V_15838	2.4	0.83 ± 0.09		345	328	0	17	yes
X_16661	2.3	0.83 ± 0.09		155	155	0	0	yes
IV_5316	2.3	0.83 ± 0.09		46064	46028	0	36	yes
IV_6089	2.2	0.83 ± 0.09		94	93	0	1	yes
X_16561	2.1	0.83 ± 0.09		696	696	0	0	yes
X_16952	2.0	0.83 ± 0.09		386	319	0	67	yes
III_4385	2.0	0.83 ± 0.09		12	8	0	4	yes
III_4580	2.0	0.83 ± 0.09		6828	5813	0	1015	yes
X_16998	2.0	0.83 ± 0.09		111	111	0	0	yes
X_16598	2.0	0.83 ± 0.09		26	26	0	0	yes
X_16902	2.0	0.83 ± 0.09		76	76	0	0	yes
I_1297	2.0	0.83 ± 0.09		204	202	0	2	yes
V_16054	2.0	0.83 ± 0.09		9890	9872	0	18	yes
I_44	1.9	0.85 ± 0.05		10	10	0	0	yes
V_15924	1.9	0.85 ± 0.05		62	62	0	0	yes
II_3022	1.9	0.85 ± 0.05		5197	5193	0	4	yes
IV_5318	1.9	0.85 ± 0.05		119	118	0	1	yes
X_16559	1.9	0.85 ± 0.05		496	496	0	0	yes

FIGURE 13.2 Output file from miRDeep2 showing novel miRNAs predicted. The table has been parsed and mature as well as hairpin miRNA sequences in columns on the right have been removed.

13.3 miRANALYZER

Another tool that identifies known miRNAs and proposes novel miRNA hairpins from RNA-seq data is miRanalyzer [4]. It also allows one to make differential comparisons of miRNA expression between two data sets. A big difference to mirDeep2 in practice is that miRanalyzer is web-based. It can also be downloaded and run locally. In either case, the user will need to cluster their RNA-seq data and reformat into the "read-count format" or "multifasta format." See Table 13.3 for how these data formats look like.

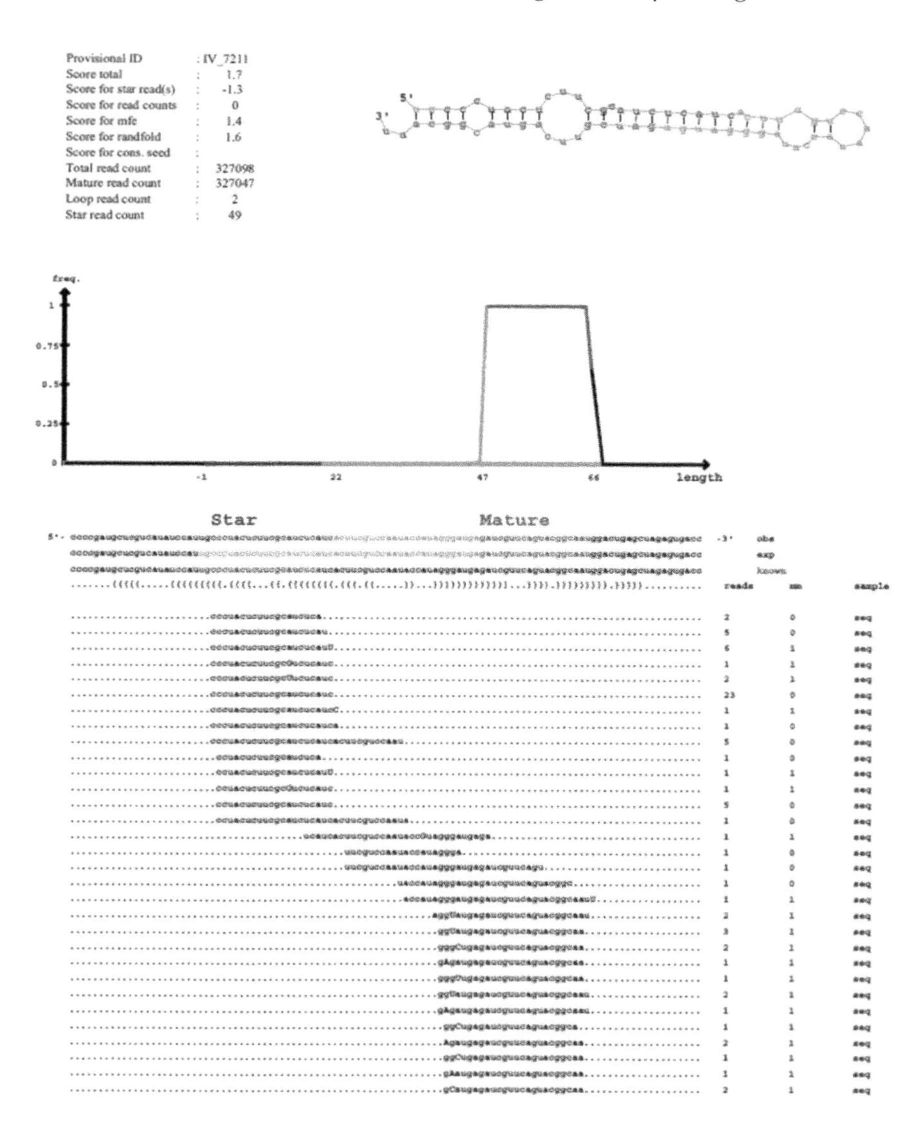

FIGURE 13.3 Graphical output of hairpin and mapped reads from miRDeep2.

Essentially these formats list the sequence and the number of reads in the data set, so it provides compression of the data set which is very important for a web-based program. A perl program is provided to perform this clustering and reformatting conversion. After that, the user is ready to go. From the web page, the user selects the organism/genome of which over 40 are currently supported including human, mice, drosophila, and *C. elegans*, enters parameters from a drop down menu such as number of

TABLE 13.3 FASTA Format, Read-Count Format, Multi-FASTA Format

FASTA Format

>gene1
ACTCTCGATCTATTT
>gene2
TCTCACGTGCGGTAAGC
>gene3
GTGATTGCATATCAT
...

READ-COUNT Format:	
ACTCTCGATCTATTT	57882
TCTCACGTGCGGTAAGC	23815
GTGATTGCATATCAT	432

MULT-IFASTA format:

>gene1 57882
ACTCTCGATCTATTT
>gene2 23815
TCTCACGTGCGGTAAGC
>gene3 432
GTGATTGCATATCAT

The table shows the differences between the formats. miRanalyzer will accept as input files data from read-count or multi-FASTA format only.

mismatches allowed, decides whether to predict novel miRNAs or detect known miRNAs and clicks the launch button. The miRanalyzer server processes and analyzes the data and provides the user with a miRanalyzerjobIDweb-link where the results are available when the job is done. Depending on the size of the data set, this could take from half a day to 1 week. The output rendered in the web page shows the parameters used and a brief summary of the results. Other sections show the number of reads mapped to known miRNAs, other RNA classes, and novel-predicted RNAs. Clicking the details box in each of these sections provides the exact miRNAs or RNA entity and reads for each. Comparison between two data sets can be made directly by using the differential expression analysis tool. Here, the user simply enters the miRanalyzerjobID of the two data sets. A tutorial and sample data sets are provided at the site of miRanalyzer. The miRanalyzer tool can be found from http://bioinfo5.ugr.es/miRanalyzer/miRanalyzer.php. The perl script for clustering and reformatting data for input to miRanalyzer can be found from http://web.bioinformatics.cicbiogune.es/microRNA/miRanalyser.php (Figure 13.4).

Queing and Execution

Analysis completed
You can bookmark this page
Download all results in plain text here

Parameters

Species:	Cel	Assembly.	Ce6
Input	mehg_ctrl.txt..	Mismatches (known):	1
Mismatches (library):	1	Mismatches (genome):	1
Score threshold:	0.9	Min. positives:	3
Type	Full analysis	Solid	no

Brief summary

unique reads:	443720	read count	18450502
filltered unique reads:	20056	filtered read count	101762
No known microRNA	274	No. known microRNA*	---
No microRNA (not miRBase)	---	No. new microRNAs	115
unique reads (after known)	395939	read count (after known):	1
unique reads (after lb)	393256	read count (after lb):	1
unique reads matched	277749	read count matched:	1
unique reads not-matched	115507	read count not-matched:	1477974

Mapping to known microRNA (miRBase 19)

Library/ Parameters	Mature	ambiguous mature	Mature-star	ambiguous mature-star	unobs. mature-star	ambiguous unobs. mature-star	hairpin	ambiguous hairpin
No.microRNA	274	7	0	0	0	0	121	3
fraction (number) of known microRNAs	74.7% (367)	---	0.0% (57)	---	---	---	54.3% (223)	---
unique reads	26882	70	0	0	0	0	768	7
fraction of unique reads	6.3%	0.017%	0.000%	0.000%	0.000%	0.000%	0.181%	0.002%
read count	2371520	488	0	0	0	0	2639	9
fraction of read count	12.9%	0.003%	0.000%	0.000%	0.000%	0.000%	0.014%	0.000%
links to detail pages	details	details	no results	no results	no results	no results	details	details

Alignment to other transcibed entities

Library/ Parameters	RefSeq_genes	Rfam
number of unique reads	58333	1196
fraction of unique reads	13.77%	0.28%
number of reads	1	3711
fraction of reads	67.55%	0.02%
Links	details	details

Predicted candidate microRNAs

No. of read clusters:	173783			
No. of checked candidates	46844			
No. new microRNAs:	115	Unique reads (read count):	629 (44440)	details
No. new microRNAs (trans filtered):	109	Unique reads (read count):	623 (4434)	details

FIGURE 13.4 Output from miRanalyzer.

13.3.1 Running miRanalyzer

1. Download the perl script for clustering and reformatting data.
2. Run the perl script to cluster and reformat your RNA-seq data.
3. Use the reformatted data as input in miRanalyzer.

13.4 miRNA TARGET ANALYSIS

miRNAs target their cognate mRNA by a not yet fully understood mechanism. However, some key aspects of target recognition are known. The complementarity of the first 7–8 nt in the miRNA to the mRNA target is important; the thermodynamic stability of the miRNA–mRNA complex is important; the position of particular GC and AU matches is important.

The nucleotides located in 5′ end of the mature miRNA in positions 2–9 have been termed the seed sequence. Complementary matches of the seed sequences in these positions with target mRNAs have been classified 7-mer-A1 (seven matching nucleotides with an adenosine at position 1), 7-mer-m8 (seven matching with a mismatch at position 8), or 8-mer (eight matching nucleotides in the seed). The target site on the mRNA has been concentrated to the mRNA 3′ untranslated region (UTR), although there is evidence that miRNAs can also target exons and 5′ UTRs. At least three different approaches have been used for miRNA target analysis. First, computational prediction methods based on scoring and statistical analysis of sequence features have been useful and historically the earliest. These prediction methods remain grounded in the original basic guidelines derived from the original lin-4 miRNA:l lin-14 mRNA pairing and later refined by identifying more miRNAs and their targets. Second, artificial intelligence methods, based on positive and negative training sets for classifying mRNAs as targets or nontargets, represent a smarter alternative to more straightforward sequence match scoring. Third, experimental methods that are considered more accurate and validated than computational methods, but suffer from the lack of throughput, have led to few examples. Early studies relied on miRNA hairpins transfected into cells followed by analysis of downregulated genes. Presently, RNA-seq methods involving cross-linking immune precipitation (CHIP) of Argonaute and other RNA binding proteins that identify the miRNA:mRNA complex *in situ* now provide the vanguard in target validation.

13.4.1 Computational Prediction Methods

Most purely computational prediction programs rely on the following guidelines. The guidelines themselves are neither necessary nor sufficient to confer targeting, but instead provide a basis to score the potential target. In practice, many different computational tools are used and consensus predictions are put forward as the best prediction.

The guidelines are

1. Complimentary matches in the "seed region" defined as position 2–7 of the miRNA
2. Compensatory complimentary matches beyond the seed region in the presence of weak seed region matches

3. Presence of adenosines flanking seed pairing sequence in mRNA
4. Multiple seed region matches of a single miRNA in the 3′UTR
5. Multiple different miRNA seed region matches in the 3′UTR
6. Overall complementary sequence matches of miRNA to the 3′UTR
7. Conservation of seed sequences across species

As the seed region is only 6–8 nucleotides, and the guidelines are vague, many computational tools suffer from lack of specificity and predictions of hundreds of targets per miRNA are common. Still, these tools are useful as a starting point for generation of new hypothesis and downstream experimental analysis. Some examples of miRNA target prediction tools follow.

TARGETSCAN (www.targetscan.org) is a general purpose web-based tool for finding predicted miRNA targets of animals. It supports a large number of species including humans, mice, flies, and nematodes. It is a curated database. Thus, predictions have been calculated with utilization of general principles of finding miRNA targets as described above [5]. miRNAs are grouped into families, so targets are listed for entire families. Output is in table format and easily exported to a spreadsheet. One advantage of this tool is that in addition to predicted mRNAs listed, a score based on conservation (Aggregate Pct) and a separate score based on sequence context (total context score) are provided to aid in evaluating confidence of predictions. As an example, miRNAs with high phylogenetic conservation would lean more heavily on Aggregate Pct score than on context. One disadvantage of the tool is the limited number of species supported and depth of support for "star" miRNA sequences. However, if you wish to make custom predictions using the target scan database, the entire database is downloadable from the site. Target scan can also be used the other way around, for example, if you have an mRNA and want to know the predicted miRNA target sites. The FAQ is elaborate and includes citations if one is interested in the precise algorithms used.

Protocol for TARGETSCAN:

1. Point your browser to TARGETSCAN (www.targetscan.org)
2. Select your species by pulling down the dialog menu and clicking your species.

3. Select your miRNA by pulling down the dialog menu and clicking your miRNA family.

4. Paste the webpage table to your spreadsheet.

DIANA-microT web server (www.microrna.gr/microT) has the unique feature that the user can input their own miRNA sequence and the server will calculate the potential targets. While it can also perform target searches for known miRNAs or vice versa, its scoring algorithm depends upon a 7, 8, or 9 nt seed match of the miRNA, G:U wobble pair in the 3′ end of the miRNA, or 6 nt match in first 9 nt of the 5′ miRNA. Conservation of the target site across 27 species is also used in the calculation. False positives are controlled by generating mock miRNAs to generate signal-to-noise ratio [6].

miRBase—miRBase contains the predictions for each miRNA from TargetScan. One simply queries the miRNA in miRBase and the output screen displays the predicted targets including the Aggregate Pct and total context score. In addition, graphic output is shown detailing the complementary match/mismatch sequences of the miRNA along with the 3′ UTR. As there is no current agreement on the best computational algorithms and/or bioinformatic tools that can be used for prediction, many laboratories use several tools to build consensus predictions. miRBase can be accessed from www.mirbase.org. An expanded explanation of miRBase will be detailed later in this chapter.

13.4.2 Artificial Intelligence Methods

Support vector machines (SVMs) are basic machine-learning tools. Features of data are mapped into high-dimensional vector space. The number of features is unlimited but can include things such as matches in the seed, mismatches in the seed, free energy of the seed or the 3′ part, and position-based sequences. Once all of the features are vectorized, different samples from a training set can be separated and a classifier built that separates best the vectors in the sample. Intuitively, this can be thought of as an optimal hyperplane that separates the vectors in the training. Next, real data are entered and the classifier that was built based on the training set is used to discriminate or separate the data. In practice, the training set can include positive examples of miRNAs that downregulate mRNAs from a microarray experiment, and negative examples from nonregulated genes from the same data set. Features are extracted from the miRNA and 3′UTR sequences of the mRNA that were downregulated. Random sequences are also sometimes used as a negative example in the training. SVMs were implemented using nine training sets with 0–51 positive examples and 0–114 negative examples with the rank importance of the first five features: nt match position 5, 5′ free energy, nt match position 6, nt match position 4, and AU matches in the 5′ part [7]. A later SVM study, using a single training set, found the important features to be seed match conservation, terminal base match, and seed 7a matching site [8]. This latter study placed their results into a web browser tool (www.mirdb.org) from where you can search for the targets of your miRNAs or miRNAs that target your gene, provided they are from human, mouse, rat, dog, or chicken [8,9].

Self-organizing maps (SOMs) are another artificial intelligent method based on an unsupervised learning algorithm that has been used to find miRNA targets. The initial learning process involves clustering data (e.g., sequences) into multidimensional space. The next mapping process involves putting new input data into the map. The mirSOM approach involved taking the 3′UTR subsequences and clustering them [10]. The result was a self-organizing map of 32 × 32 = 964 neurons, with each neuron containing none, one or more of 1.8 million 22 nt substrings from the 3′UTR of known *C. elegans* genes. Next, the miRNAs were mapped onto the SOM and the genes corresponding to the 3′UTR sequences in the neuron were considered candidate targets. mirSOM performed well against most other tools with high sensitivity and vastly improved specificity. Unfortunately, it currently supports only *C. elegans* data. The mirSOM interface allows the user to enter an miRNA and the predicted mRNAs are returned as output. mirSOM can be accessed from www.oppi.uef.fi/bioinformatics/mirsom/.

13.4.3 Experimental Support-Based Methods

Biologists are more comfortable with computational predictions if there is experimental support. Fortunately, during the past years, many efforts have been made toward this end. Most experimental methods are limited by through-put, however they have value in providing more certain miRNA:mRNA target interactions. Some of the experimental methods are now of sufficiently high throughput that a database of their own for results are justified. More recently, a curated database covering miRNAs with experimental support from the literature has proved highly useful. Below we describe several database sources that contain miRNA::mRNA targets, both confirmed and predicted, with support based on experimental data.

mirWIP—As an alternative to pure computational prediction, it is possible to obtain a list of miRNA–mRNA duplexes obtained from immunoprecipitation experiments that pull out the RISC complex from *C. elegans* [11,12]. This knowledge was first used to provide a list of experimentally verified miRNA:mRNA target complexes and secondly to develop an improved scoring criteria for miRNA target prediction. Initial target prediction in mirWIP relied on minimal free energy, phylogenetic conservation, and seed pairing. mirWIP then used the experimental immunoprecipitation data to refine their scoring algorithm which now includes 5′ seed match features, structural accessibility, and binding site energy. mirWIP can be accessed from www.mirwip.org.

TarBase is a curated database of miRNA targets with experimental support. TarBase curates from the literature in a semiautomated fashion suing a text-mining-assisted curation pipeline [13]. The database contains 65,814 experimentally validated miRNA–gene interactions. The experimental data are sourced from miRNA gene-specific methods such as reporter genes, qRT-PCR, and Western blotting, to more high-throughput methods such as microarrays, and RNA-seq methods such as HITS-CLIP and PAR-CLIP, and Degradome-seq. In surprising turnaround, computational verification of the experimental results is available through DIANA microTmiRNA target scores. Data and access to a very friendly user interface from TarBase can be found at the following website (www.microrna.gr/tarbase).

miRTarBase is another curated miRNA-target database curated from the literature. With nearly 2000 articles curated, it has 3576 experimentally verified miRNA target interactions from 17 species including humans [14]. Compared to TarBase, it does not contain miRNA–target interactions from RNA-seq studies and thus might be preferred by those interested in data from single-gene-derived validation studies. miRTarBase can be accessed from the following website (www.miRTarBase.mbc.nctu.edu.tw/).

13.5 miRNA-SEQ AND mRNA-SEQ DATA INTEGRATION

As a goal of many miRNA-seq studies, one would like to know not only which miRNAs are regulated in a specific tissue sample, but also which mRNAs those miRNAs regulate. This problem has been under study for many years even prior to NGS. For individual miRNAs, as shown above, a straightforward step would be to select an miRNA from a list of dysregulated miRNAs to focus on and computationally predict the mRNA target. Alternatively, many laboratories take it a step further by selecting a specific miRNA and isolating its functional role by transfecting or transforming cells with the miRNA hairpin mimic and then confirming its regulated mRNAs by qRT-PCR, microarray, or RNA-seq methods. This can also be done the other way around, and miRNA knockouts or knockdowns can be performed and candidate mRNA targets assayed for their expression. While these laboratory methods are laborious, they do demonstrate experimental validation of an miRNA and its target.

A typical miRNA-seq study may generate tens of dysregulated miRNAs. The equivalent RNA-seq study on the same tissues may yield hundreds of dysregulated mRNAs. How to integrate the results is no trivial task. A simple approach has been to correlate the expression levels of miRNAs in a tissue or cell type with the mRNA levels in the same tissues. A more sophisticated approach is to add transcription factor target prediction and build gene networks based on this data [15]. One approach representing this has been to build miRNA-transcription factor feed-forward loops [16]. In this solution, miRNA-transcription factor-mRNA networks are built via transcription factor databases and miRNA-target prediction tools. These networks are then integrated with miRNA and mRNA expression data from gene expression databases such as the Gene expression omnibus (GEO) and ArrayExpress. Target enrichment data are then statistically scored where the best scores are given for miRNAs-transcription factor-mRNA networks that have support from the gene expression databases.

miRNA–mRNA data integration continues to be challenging. An underlying factor is that miRNA and mRNA interactions occur in 4D, that is, they are dependent upon not only spatial considerations, but also in time during development and aging. It is not always obvious whether a 2-fold increase in miRNA from an miRNA-seq experiment means that twice as many cells express the miRNA or that all the cells express it twice as abundantly. If the miRNA is expressed, are the other factors present that would allow it to target and regulate? Currently, there is much work to be done in this area and seemingly no easy answers.

13.6 SMALL RNA DATABASES AND RESOURCES

13.6.1 RNA-seq Reads of miRNAs in miRBase

miRBase is the curated miRNA database for research purposes. It contains sequences, structures, putative targets, and references for many species [17]. miRBase allows you to search for miRNA sequences and annotation based on miRNA name or keyword. It currently contains over

20,000 miRNA entries (Release19, August 2012). One can browse miRNAs by species. For example, the current version in miRBase (release 20) has 1872 human miRNA precursors and 2578 mature sequences. The download section allows one to download all miRNA sequences in the database or just the precursors or just the mature sequences. It also has downloadable files for miRNA families. A new feature recently added was curated data on the reads supporting the existence of each miRNA [18]. Annotating and maintaining the number of experimentally verified reads for miRNAs continues to be a challenge, however miRBase appears to be doing an exceptional job in the face of massive amounts of new data [19]. A typical example of a miRBase entry cel-mir-124 is shown in Figure 13.5. This entry shows the accession number, description, gene family, community annotation, the stem loop with matching, loop, and bulge regions, number of deep sequencing reads that support the annotation, location of the deep sequencing reads, and genome context which is essentially the precise coordinates and location relative to transcripts.

Further results provide experimental evidence for the miRNA, validated targets, and predicted targets. The experimental evidence can be from cloning, Northern blots, sequencing, or CLIPseq. The validated targets come from a link with TARBASE, and the predicted targets are from MICRORNA.ORG, RNA22-CEL, or TARGETSCAN-WORM. Finally, a section provides the most important references relevant for the specific miRNA.

From the view of the Stem-loop, it is possible to obtain the precise number of reads for the miRNAs in the stem loop. Clicking the number in front of reads opens another view with deep sequencing reads for the stem–loop sequence shown in Figure 13.6. Here, one can view the sequence of the reads and the exact number of reads/counts for the specific sequence in both absolute values or counts and in normalized RPM (mean number of reads per million) units. Another advantage of this view is that one can see the sequenced iso-mirs and their contribution to the overall counts. A further view provides the experiment accession, and read count from where the data is derived.

UCSC genome browser is a versatile all-purpose browser to find features of a sequence at the genome level using annotation tracks. Each track lists a specific annotation based on biological experiments and divided into categories. There are hundreds of possible annotation tracks for any sequence and therefore it is interactive allowing the user to click on various annotation tracks. For RNA-seq data, the user can also view the reads from sequencing experiments across a defined region of the genome. Data are

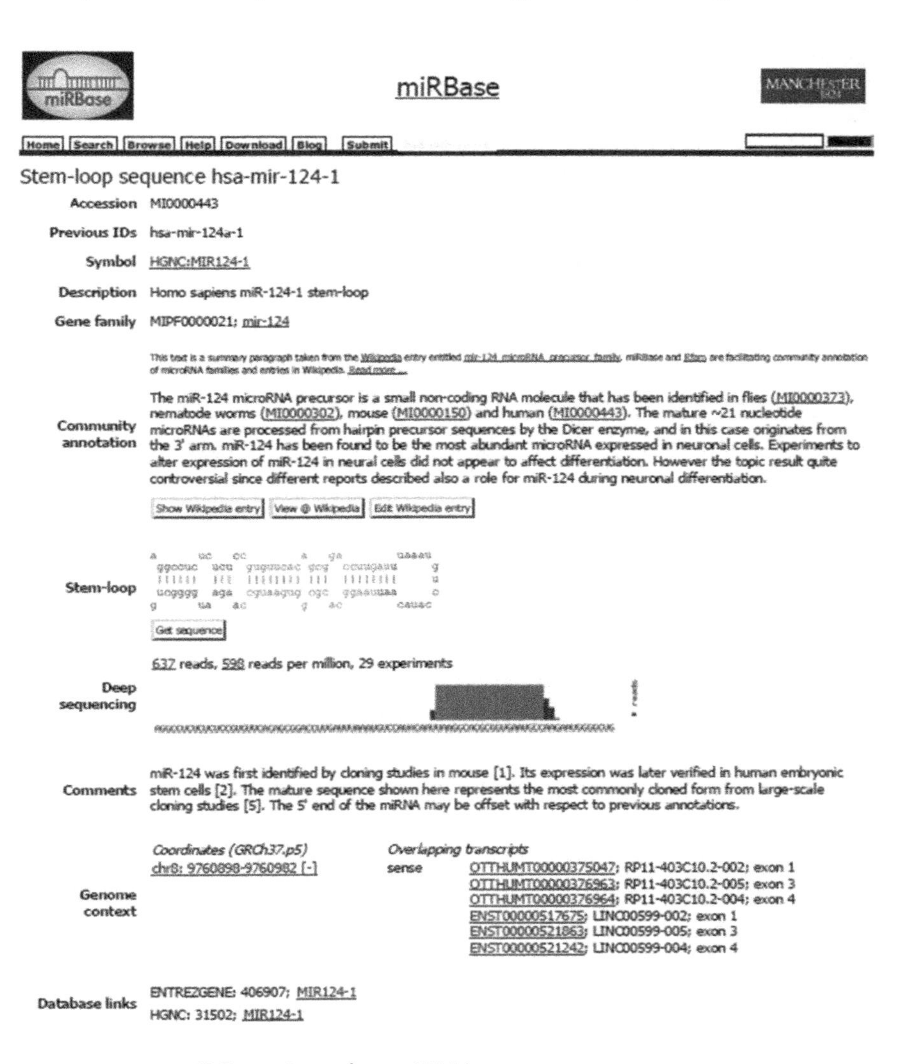

FIGURE 13.5 miRBase view of an miRNA.

curated from different experiments including the Encode project. UCSC genome browser can be accessed from http://genome.ucsc.edu/.

13.6.2 Expression Atlas of miRNAs

microRNA.org at the website under the same name (microRNA.org) has essentially two tools [20]. The first lists results of computational predictions for miRNA:mRNA target sites based on an intelligent support vector regression algorithm trained from transcription experiments. The user can search with miRNA for targets or vice versa. A unique and powerful

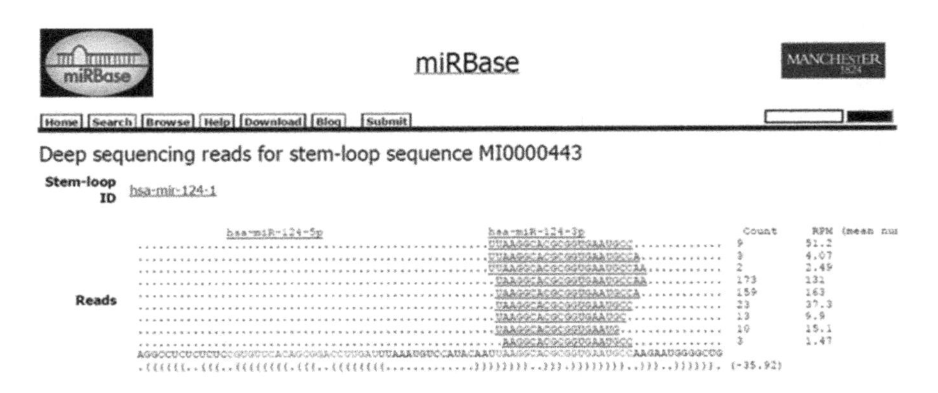

FIGURE 13.6 miRBase entry of hsa-mirl24-1 RNA-seq reads curated from experiments.

feature allows the user to access the expression profile of miRNAs that were generated from construction of small RNA libraries and sequencing >300,000 clones from 256 small RNA libraries from 26 organ systems or cell lines of humans, rat, and mice. The expression pattern of an miRNA can be viewed in heatmap, bargraph, or 3D bargraph styles.

piRNABank contains searchable piwi sequences, their location, maps, and tools for analysis [21]. The resource supports piRNA from human, mouse, rat, drosophila, zebra fish, and platypus. There are currently over 20,000 nonoverlapping human piRNA sequences in the database. A useful utility is the ability to view and download piRNA sequences in a cluster. So if you are interested in all piRNAs expressed in a specific genome location, you can get all hundreds or thousands of piRNA sequences downloaded in FASTA format. piwiDB can be found at the following website (pirnabank.ibab.ac.in/index.shtml).

Rfam is one of the oldest RNA databases existing for over 10 years [22]. The database classifies RNA based on families and represents them via alignments, secondary structures, and covariance models. It contains not only noncoding RNA genes, but also catalytic RNAs. The covariance models are used to model both RNA sequence and structure. The database allows one to not only retrieve RNA data, but also to find matches to input RNA sequences and view annotation and alignments. The most recent version of Rfam (v11.0) contained 2208 RNA families. Rfam can be found at http://rfam.sanger.ac.uk/.

miRGator [23] can be found at mirgator.kobic.re.kr. It is an RNA database portal connected to NGS, expression, and mRNA target data.

13.6.3 Database for CLIP-seq and Degradome-seq Data

starBase represents a database that has collected and analyzed 21 Argonaute or TNRC6 CLIP-seq and 10 Degradome-seq data from six organisms include human, mouse, *C. elegans*, *Arabidopsis thaliana*, *Oryza sativa* (rice) and *Vitis vinifera* (wine grape) [24]. The database is interactive and displays the genomic landscape of Argonaute-binding sites and miRNA cleavage sites. Combining data from the CLIP-seq and Degradome-seq data sets with computational predictions produced approximately 66,000 miRNA-target regulatory relationships, a vast increase of current knowledge. starBase is searchable, so you can enter a seed sequence and locate a CLIP-seq target via the CLIPSearch server or enter a short RNA sequence and locate a Degradome-seq sequence via the Degradome Search server. It has been recently updated [25]. starBase can be accessed from http://starbase.sysu.edu.cn/.

13.6.4 Databases for miRNAs and Disease

miRò is a web-based interface that helps users to find links between miRNAs and diseases [26]. The database integrates data from miRNAs with mRNA target predictions (TargetScan, PicTar, miRanda) and experimental validations (miRecords). Genes that associate with the miRNAs are then linked to Gene Ontology and Genetic Association Database. The database then essentially links your miRNA to a human disease. Four types of queries can be formed: retrieve information about an miRNA, gene, ontology term, disease, or tissue; find an association between an miRNA and a disease, or the other way around, find all miRNAs associated with a disease; test new miRNA-target pairs for associations; and perform an advanced query by choosing a subject and specifying constraints to the subject to limit output. For example, query with "Parkinson's disease" will return a list of 7658 entries with the miRNA, gene it targets, and whether the target has been experimentally validated or predicted and which programs predicted it. miRò can be accessed from http://ferrolab.dmi.unict.it/miro/index.php.

miRdSNP is a database for those wishing to associate miRNAs with their target sites in the 3′UTR that have disease-associated single-nucleotide polymorphisms (SNPs) located nearby [27]. The database currently has 786 disease-associated SNPs and 204 diseases. The tool is especially useful for those who want to find a potential SNP associated with an miRNA. miRdSNP can be accessed from mirdsnp.ccr.buffalo.edu.

13.6.5 General Databases for the Research Community and Resources

RNAcentral is an effort aimed toward a federated database of RNA sequences. As the number of small noncoding RNA families grows and research into their function accompanies this growth, databases become built to accommodate the RNA sequences and annotations concerning their biological function. This has resulted in highly expert and specialized databases for many individual RNA families (miRBase, RFAM, starBase, etc.). An advantage of this trend is domain-specific knowledge and ability to keep up with rapidly evolving new information that impacts the field. The downside is the fragmentation of information regarding RNA sequences and different interfaces and database models an RNA biology researcher must negotiate. Therefore, many of these individual RNA database managers have envisioned and are working toward a federated database of RNA sequences [28]. The federated approach allows each database to maintain their own identity, management, administration, and interact with the user, but also takes advantage of a centralized database and web portal that can be a resource to the user community for RNA sequence accession, storage, and representation. Such an approach has been successful for the protein field (e.g., InterPro database). The project has received funding and is currently being built at EMBL—European Bioinformatics Institute.

13.6.6 miRNAblog

The miRNAblog (mirnablog.com) is a centralized resource to obtain updates on miRNA studies, find out where conferences will be held, list jobs, find jobs, and solve common problems in the miRNA field. It is supported by advertisements and the scientific community of miRNA researchers. Academic research reports as well as industry news is easily found and discussed in the blog.

It is becoming abundantly clear that most miRNA portals are striving to incorporate RNA-seq data retrieval, representation, and data analysis tools into their systems. While the scale of this effort is ambitious and requires top-level engineering as well as computational and biological support, the fruit from this venture is eagerly anticipated and will be rapidly and voraciously consumed by downstream users in the near future. A summary of all the sites and resources presented here can be found in Table 13.4.

TABLE 13.4 Resources for miRNA-seq Analysis

Resource	Type	Description	Address
miRBase	Database	Database of miRNAs, RNA-seq reads shown for miRNA sequences	mirbase.org
RFAM	Database	Database of RNA families	rfam.sanger.ac.uk
UCSC	Database	Database and genome browser	genome.ucsc.edu/
piRNABank	Database	Database of piwiRNA sequences	pirnabank.ibab.ac.in/index.shtml
starBase	Database	Database of CLIP-seq and degradome data	starbase.sysu.edu.cn
microRNA.org	Database	Database of miRNA libraries and expression profiles	microRNA.org
miRò	Database	Database of miRNA linked to targets and disease	ferrolab.dmi.unict.it/miro/index.php
miRdSNP	Database	Database of miRNAs, miRNA targets, associated SNPs and diseases	mirdsnp.ccr.buffalo.edu
miRDeep2	Analysis	Analysis of small RNA-Seq data: Discovery, annotation, display	mdc-berlin.de/8551903/en/research/research_teams/systems_biology_of_gene_regulatory_elements/projects/miRDeep
miRanalyzer	Analysis	Analysis of small RNA-Seq data: Discovery, annotation, display	http://bioinfo5.ugr.es/miRanalyzer/miRanalyzer.php
TargetScan	Analysis	miRNA Target tool	targetscan.org
mir-WIP	Analysis	miRNA Target tool based on improved predictions from Immunoprecipitation data	146.189.76.171/query.php
mirSOM	Analysis	miRNA Target tool based on self-organizing maps	www.oppi.uef.fi/bioinformatics/mirsom/
mirTarBase	Analysis	miRNA Target tool based on curated literature	miRTarBase.mbc.nctu.edu.tw
TarBase	Analysis	miRNA Target tool based on experimental evidence	microrna.gr/tarbase
DIANA-microT	Analysis	miRNA Target tool that allows you to enter your own miRNA and returns potential targets	ferrolab.dmi.unict.it/miro/index.php
microRNA blog	General	News, meetings, announcements, jobs	mirnablog.com

13.7 SUMMARY

There currently exists a large set of readily available and easy to use tools for miRNAdata analysis. These tools include those for handling raw NGS reads as well as downstream analyses. In addition, several databases, highlighted by miRBase, provide the scientific community with a rich source of curated data. Some tools require familiarity with a command line environment; however, a growing number of analysis tools are web-based. Taken together, these tools provide a powerful set of implementations that allows one to discover, annotate, visualize, and identify differences in miRNA data sets. While analysis tools for analysis of non-miRNA classes are lagging, more effort will been seen in this area in the near future. A practical approach to miRNA and small noncoding RNA sequence analysis is indeed readily possible in the current environment.

REFERENCES

1. Li Y., Zhang Z., Liu F. et al. Performance comparison and evaluation of software tools for microRNA deep-sequencing data analysis. *Nucleic Acids Research* 40(10):4298–4305, 2012.
2. Williamson V., Kim A., Xie B. et al. Detecting miRNAs in deep-sequencing data: A software performance comparison and evaluation. *Briefings Bioinformatics* 14(1):36–45, 2013.
3. Friedländer M.R., Mackowiak S.D., Li N., Chen W. et al. miRDeep2 accurately identifies known and hundreds of novel microRNA genes in seven animal clades. *Nucleic Acids Research* 40(1):37–52, 2012.
4. Hackenberg M., Sturm M., Langenberger D. et al. miRanalyzer: A microRNA detection and analysis tool for next-generation sequencing experiments. *Nucleic Acids Research* 37(Web Server issue):W68–W76, 2009.
5. Lewis B.P., Burge C.B., and Bartel D.P. Conserved seed pairing, often flanked by adenosines, indicates that thousands of human genes are microRNA targets. *Cell* 120(1):15–20, 2005.
6. Maragkakis M., Reczko M., Simossis V.A. et al. DIANA-microT web server: Elucidating microRNA functions through target prediction. *Nucleic Acids Research* 37(Web Server issue):W273–W276, 2009.
7. Kim S.K., Nam J.W., Rhee J.K. et al. miTarget: microRNA target gene prediction using a support vector machine. *BMC Bioinformatics* 7:411, 2006.
8. Wang X. and El Naqa I.M. Prediction of both conserved and nonconserved microRNA targets in animals. *Bioinformatics* 24(3):325–332, 2008.
9. Wang X. miRDB: A microRNA target prediction and functional annotation database with a wiki interface. *RNA* 14(6):1012–1017, 2008.
10. Heikkinen L., Kolehmainen M., and Wong G. Prediction of microRNA targets in *Caenorhabditis elegans* using a self-organizing map. *Bioinformatics* 27(9):1247–1254, 2011.

11. Zhang L., Ding L., Cheung T.H. et al. Systematic identification of *C. elegans* miRISC proteins, miRNAs, and mRNA targets by their interactions with GW182 proteins AIN-1 and AIN-2. *Molecular Cell* 28(4):598–613, 2007.
12. Hammell M., Long D., Zhang L. et al. mirWIP: MicroRNA target prediction based on microRNA-containing ribonucleoprotein-enriched transcripts. *Nature Methods* 5(9):813–819, 2008.
13. Vergoulis T., Vlachos I.S., Alexiou P. et al. TarBase 6.0: Capturing the exponential growth of miRNA targets with experimental support. *Nucleic Acids Research* 40(Database issue):D222–D229, 2012.
14. Hsu S.D., Lin F.M., Wu W.Y. et al. miRTarBase: A database curates experimentally validated microRNA-target interactions. *Nucleic Acids Research* 39(Database issue):D163–D169, 2011.
15. Mestdagh P., Lefever S., Pattyn F. et al. The microRNA body map: Dissecting microRNA function through integrative genomics. *Nucleic Acids Research* 39(20):e136, 2011.
16. Yan Z., Shah P.K., Amin S.B. et al. Integrative analysis of gene and miRNA expression profiles with transcription factor-miRNA feed-forward loops identifies regulators in human cancers. *Nucleic Acids Research* 40(17):e135, 2012.
17. Griffiths-Jones S., Saini H.K., van Dongen S. et al. miRBase: Tools for microRNA genomics. *Nucleic Acids Research* 36(Database issue):D154–D158, 2008.
18. Kozomara A. and Griffiths-Jones S. miRBase: Integrating microRNA annotation and deep-sequencing data. *Nucleic Acids Research* 39(Database Issue):D152–D157, 2011.
19. Kozomara A. and Griffiths-Jones S. miRBase: Annotating high confidence microRNAs using deep sequencing data. *Nucleic Acids Research* 42(1):D68–D73, 2014.
20. Betel D., Wilson M., Gabow A. et al. The microRNA.org resource: Targets and expression. *Nucleic Acids Research* 36(Database issue):D149–D153, 2008.
21. Lakshmi S. and Agrawal S. piRNABank: A web resource on classified and clustered Piwi-interacting RNAs. *Nucleic Acids Research* 36(Database issue):D173–D177, 2008.
22. Burge S.W., Daub J., Eberhardt R. et al. Rfam 11.0: 10 years of RNA families. *Nucleic Acids Research* 41(Database issue):D226–D232, 2013.
23. Cho S., Jang I., Jun Y., Yoon et al. MiRGator v3.0: A microRNA portal for deep sequencing, expression profiling and mRNA targeting. *Nucleic Acids Research* 41(Database issue):D252–D257, 2013.
24. Yang J.H., Li J.H., Shao P. et al. starBase: A database for exploring microRNA–mRNA interaction maps from Argonaute CLIP-seq and Degradome-seq data. *Nucleic Acids Research* 39(Database issue):D202–D209, 2011.
25. Li J.H., Liu S., Zhou H. et al. starBase v2.0: Decoding miRNA–ceRNA, miRNA–ncRNA and protein–RNA interaction networks from large-scale CLIP-seq data. *Nucleic Acids Research* 42(1):D92–D97, 2014.
26. Laganà A., Forte S., Giudice A. et al. miRò: A miRNA knowledge base. Database (Oxford):bap008 2009.

27. Bruno A.E., Li L., Kalabus J.L. et al. miRdSNP: A database of disease-associated SNPs and microRNA target sites on 3′UTRs of human genes. *BMC Genomics* 13:44, 2012.
28. Bateman A., Agrawal S., Birney E. et al. RNAcentral: A vision for an international database of RNA sequences. *RNA* 17(11):1941–1946, 2011.

Index

C

D

H

I

J

K

L

M

N

O

P

S

T

Accessing the E-book edition

Using the VitalSource® ebook

Access to the VitalBook™ ebook accompanying this book is via VitalSource® Bookshelf – an ebook reader which allows you to make and share notes and highlights on your ebooks and search across all of the ebooks that you hold on your VitalSource Bookshelf. You can access the ebook online or offline on your smartphone, tablet or PC/Mac and your notes and highlights will automatically stay in sync no matter where you make them.

1. **Create a VitalSource Bookshelf account at** ***https://online.vitalsource.com/user/new*** or log into your existing account if you already have one.
2. **Redeem the code provided in the panel below to get online access to the ebook.** Log in to Bookshelf and select **Redeem** at the top right of the screen. Enter the redemption code shown on the scratch-off panel below in the **Redeem Code** pop-up and press **Redeem**. Once the code has been redeemed your ebook will download and appear in your library.

No returns if this code has been revealed.

DOWNLOAD AND READ OFFLINE

To use your ebook offline, download BookShelf to your PC, Mac, iOS device, Android device or Kindle Fire, and log in to your Bookshelf account to access your ebook:

On your PC/Mac

Go to ***https://support.vitalsource.com/hc/en-us*** and follow the instructions to download the free **VitalSource Bookshelf** app to your PC or Mac and log into your Bookshelf account.

On your iPhone/iPod Touch/iPad

Download the free **VitalSource Bookshelf** App available via the iTunes App Store and log into your Bookshelf account. You can find more information at ***https://support.vitalsource.com/hc/en-us/categories/200134217-Bookshelf-for-iOS***

On your Android™ smartphone or tablet

Download the free **VitalSource Bookshelf** App available via Google Play and log into your Bookshelf account. You can find more information at ***https://support.vitalsource.com/hc/en-us/categories/200139976-Bookshelf-for-Android-and-Kindle-Fire***

On your Kindle Fire

Download the free **VitalSource Bookshelf** App available from Amazon and log into your Bookshelf account. You can find more information at ***https://support.vitalsource.com/hc/en-us/categories/200139976-Bookshelf-for-Android-and-Kindle-Fire***

N.B. The code in the scratch-off panel can only be used once. When you have created a Bookshelf account and redeemed the code you will be able to access the ebook online or offline on your smartphone, tablet or PC/Mac.

SUPPORT

If you have any questions about downloading Bookshelf, creating your account, or accessing and using your ebook edition, please visit ***http://support.vitalsource.com/***